WORKING GUI
RESERVOIR ENGINEERING

WORKING GUIDE TO
RESERVOIR
ENGINEERING

___

WILLIAM C. LYONS

ELSEVIER

AMSTERDAM • BOSTON • HEIDELBERG • LONDON
NEW YORK • OXFORD • PARIS • SAN DIEGO
SAN FRANCISCO • SINGAPORE • SYDNEY • TOKYO
Gulf Publishing is an imprint of Elsevier

Gulf Professional Publishing is an imprint of Elsevier
30 Corporate Drive, Suite 400, Burlington, MA 01803, USA,
The Boulevard, Langford Lane, Oxford OX5 1GB

First edition 2010

Notice
No responsibility is assumed by the publisher for any injury and/or damage to persons or
property as a matter of products liability, negligence or otherwise, or from any use or
operation of any methods, products, instructions or ideas contained in the material herein.
Because of rapid advances in the medical sciences, in particular, independent verification of
diagnoses and drug dosages should be made.

Library of Congress Cataloging in Publication Data
A catalog record for this book is available from the Library of Congress

British Library Cataloguing in Publication Data
A catalogue record for this book is available from the British Library

ISBN: 978-1-85617-824-2

For information on all Elsevier publications visit
our website at *elsevierdirect.com*

Typeset by: diacriTech, India

Printed and bound in United States of America
Transferred to Digital Printing in 2014

Contents

Contents

Full Contents

Chapter 1

BASIC PRINCIPLES, DEFINITIONS, AND DATA

Chapter 2

FORMATION EVALUATION

Chapter 3

MECHANISMS & RECOVERY OF HYDROCARBONS BY NATURAL MEANS

Chapter 4

FLUID MOVEMENT IN WATERFLOODED RESERVOIRS

Chapter 5

ENHANCED OIL RECOVERY METHODS

1

Basic Principles, Definitions, and Data

1.1 RESERVOIR FLUIDS

1.1.1 Oil and Gas

Reservoir oil may be saturated with gas, the degree of saturation being a function, among others, of reservoir pressure and temperature. If the reservoir oil has dissolved in it all the gas it is capable of holding under given conditions, it is referred to as saturated oil. The excess gas is then present in the form of a free gas cap. If there is less gas present in the reservoir than the amount that may be dissolved in oil under conditions of reservoir pressure and temperature, the oil is then termed undersaturated. The pressure at which the gas begins to come out of solution is called the saturation pressure or the bubble-point pressure. In the case of saturated oil, the saturation pressure equals the reservoir pressure and the gas begins coming out of solution as soon as the reservoir pressure begins to decrease. In the case of undersaturated oil, the gas does not start coming out of solution until the reservoir pressure drops to the level of saturation pressure.

Apart from its function as one of the propulsive forces, causing the flow of oil through the reservoir, the dissolved gas has other important effects on recovery of oil. As the gas comes out of solution the viscosity of oil increases and its gravity decreases. This makes more difficult the flow of oil through the reservoir toward the wellbore. Thus the need is quite apparent for production practices tending to conserve the reservoir pressure and retard the evolution of the dissolved gas. Figure 1.1 shows the effect of the dissolved gas on viscosity and gravity of a typical crude oil.

The dissolved gas also has an important effect on the volume of the produced oil. As the gas comes out of solution the oil shrinks so that the liquid oil at surface conditions will occupy less volume than the gas-saturated oil occupied in the reservoir. The number of barrels of reservoir oil at reservoir pressure and temperature which will yield one barrel of stock tank oil at 60°F and atmospheric pressure is referred to as the formation volume factor or reservoir volume factor. Formation volume factors are described in a subsequent section. The solution gas–oil ratio is the number of standard cubic feet of gas per barrel of stock tank oil.

Physical properties of reservoir fluids are determined in the laboratory, either from bottomhole samples or from recombined surface separator

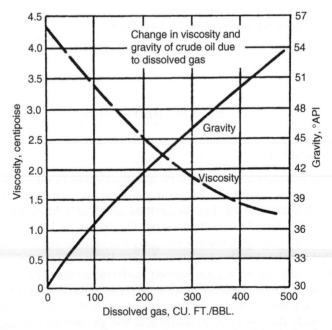

FIGURE 1.1   Change in viscosity and gravity of crude oil due to dissolved gas.

samples. Frequently, however, this information is not available. In such cases, charts such as those developed by M.B. Standing and reproduced as Figures 1.2, 1.3, 1.4, and 1.5 have been used to determine the data needed [1, 2]. The correlations on which the charts are based present bubble-point pressures, formation volume factors of bubble-point liquids, formation volume factors of gas plus liquid phases, and, density of a bubble-point liquid as empirical functions of gas-oil ratio, gas gravity, oil gravity, pressure, and temperature.

Traditionally, most estimates of PVT properties were obtained by using charts and graphs of empirically derived data. With the development of programmable calculators, graphical data are being replaced by mathematical expressions suitable for computer use. In a later section, the use of such programs for estimating PVT properties will be presented. In the initial sections, the presentation of graphical data will be instructive to gaining a better understanding of the effect of certain variables.

1.1.2 Water

Regardless of whether a reservoir yields pipeline oil, water in the form commonly referred to as interstitial or connate is present in the reservoir in pores small enough to hold it by capillary forces.

The theory that this water was not displaced by the migration of oil into a water-bearing horizon is generally accepted as explanation of its presence.

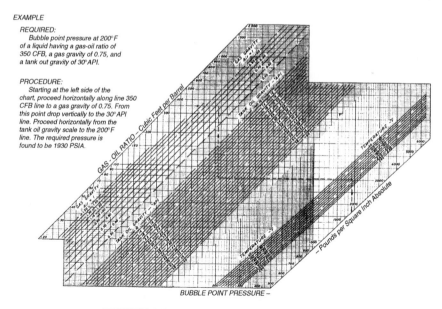

FIGURE 1.2   Bubble point pressure graph [1, 2].

EXAMPLE

REQUIRED:
  Formation volume at 200°F of a bubble point liquid having a gas-oil ratio of 350 CFB, a gas gravity of 0.75, and a tank oil gravity of 30°API.

PROCEDURE:
  Starting at the left side of the chart, proceed horizontally along the 350 CFB line to a gas gravity of 0.75. From this point drop vertically to the 30°API line. Proceed horizontally from the tank oil gravity scale to the 200°F line. The required formation volume is found to be 1.22 barrel per barrel of tank oil.

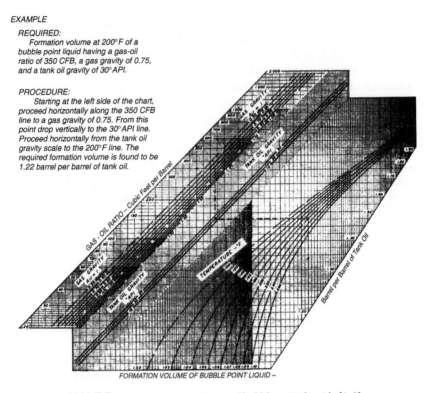

FORMATION VOLUME OF BUBBLE POINT LIQUID –

FIGURE 1.3   Formation volume of bubble point liquids [1, 2].

The amount of the interstitial water is usually inversely proportional to the permeability of the reservoir. The interstitial water content of oil-producing reservoirs often ranges from 10% to 40% of saturation.

Consideration of interstitial water content is of particular importance in reservoir studies, in estimates of crude oil reserves and in interpretation of electrical logs.

1.1.3 Fluid Viscosities

Gas Viscosity   Viscosities of natural gases are affected by pressure, temperature, and composition. The viscosity of a specific natural gas can be measured in the laboratory, but common practice is to use available empirical data such as those shown in Figures 1.6 and 1.7. Additional data are given in the *Handbook of Natural Gas Engineering* [3]. Contrary to the case for liquids, the viscosity of a gas at low pressures increases as the temperature is raised. At high pressures, gas viscosity decreases as the temperature is raised. At intermediate pressure, gas viscosity may decrease as temperature is raised and then increase with further increase in temperature.

EXAMPLE

REQUIRED:
  Formation volume of the gas plus
liquid phases of a 1500 CFB mixture,
gas gravity= 0.80, tank oil gravity= 40° API,
at 200° F and 1000 PSIA.

PROCEDURE:
  Starting at the left side of the chart
proceed horizontally along the 1500 CFB line
to the 0.80 gas gravity line. From this point
drop vertically to the 40° API line. Proceed
horizontally to 200° F and from that point
drop to line 1000 PSIA pressure line. The
required formation volume is found to be
5.0 barrels per barrel of tank oil.

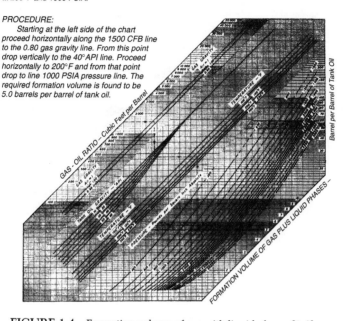

FIGURE 1.4 Formation volume of gas with liquid phases [1, 2].

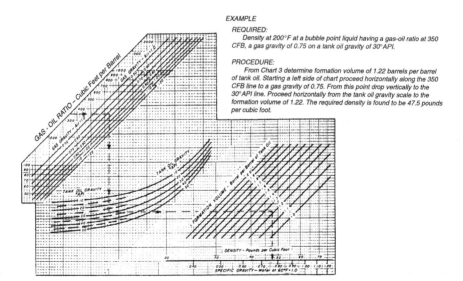

EXAMPLE

REQUIRED:
  Density at 200°F at a bubble point liquid having a gas-oil ratio at 350
CFB, a gas gravity of 0.75 on a tank oil gravity of 30° API.

PROCEDURE:
  From Chart 3 determine formation volume of 1.22 barrels per barrel
of tank oil. Starting a left side of chart proceed horizontally along the 350
CFB line to a gas gravity of 0.75. From this point drop vertically to the
30° API line. Proceed horizontally from the tank oil gravity scale to the
formation volume of 1.22. The required density is found to be 47.5 pounds
per cubic foot.

FIGURE 1.5 Density and specific gravity of mixtures [1, 2].

WORKING GUIDE TO RESERVOIR ENGINEERING

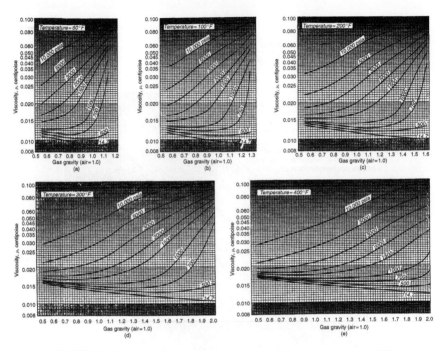

FIGURE 1.6   Gas viscosity versus gravity at different temperatures [3].

Oil Viscosity   The viscosity of crude oil is affected by pressure, temperature, and most importantly, by the amount of gas in solution. Figure 1.8 shows the effect of pressure on viscosities of several crude oils at their respective reservoir temperatures [4]. Below the bubble-point, viscosity decreases with increasing pressure because of the thinning effect of gas going into solution. Above the bubble-point, viscosity increases with increasing pressure because of compression of the liquid. If a crude oil is undersaturated at the original reservoir pressure, viscosity will decrease slightly as the reservoir pressure decreases. A minimum viscosity will occur at the saturation pressure. At pressures below the bubble-point, evolution of gas from solution will increase the density and viscosity of the crude oil as the reservoir pressure is decreased further.

Viscosities of hydrocarbon liquids decrease with increasing temperature as indicated in Figure 1.9 for gas-free reservoir crudes [5]. In cases where only the API gravity of the stock tank oil and reservoir temperature are known, Figure 1.9 can be used to estimate dead oil viscosity at atmospheric pressure. However, a more accurate answer can be obtained easily in the laboratory by simply measuring viscosity of the dead oil with a viscometer at reservoir temperature.

With the dead oil viscosity at atmospheric pressure and reservoir temperature (either measured or obtained from Figure 1.9), the effect of solution

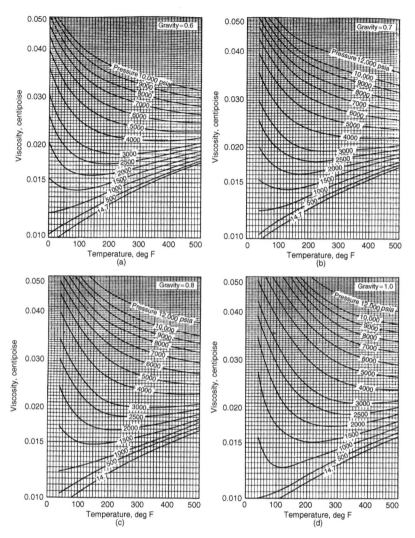

FIGURE 1.7 Gas viscosity versus temperature at different gravities [3].

gas can be estimated with the aid of Figure 1.10 [6]. The gas-free viscosity and solution gas-oil ratio are entered to obtain viscosity of the gas-saturated crude at the bubble-point pressure. This figure accounts for the decrease in viscosity caused by gas going into solution as pressure is increased form atmospheric to the saturation pressure.

If the pressure is above the bubble-point pressure, crude oil viscosity in the reservoir can be estimated with Figure 1.11 [5]. This figure shows the increase in liquid viscosity due to compression of the liquid at pressures higher than the saturation pressure. Viscosity of the crude can be estimated

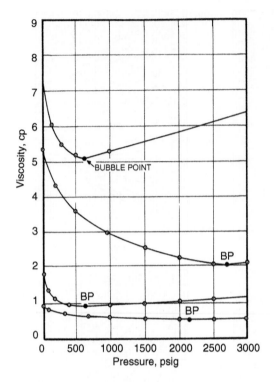

FIGURE 1.8   Effect of pressure on crude oil viscosities [4].

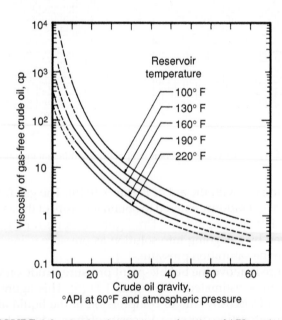

FIGURE 1.9   Crude oil viscosity as a function of API gravity [5].

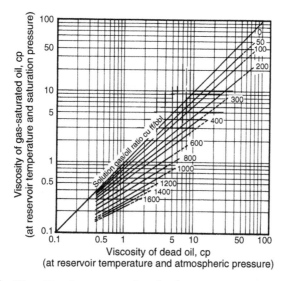

FIGURE 1.10 Viscosities of gas-saturated crude oils at reservoir temperature and bubble-point pressure [6].

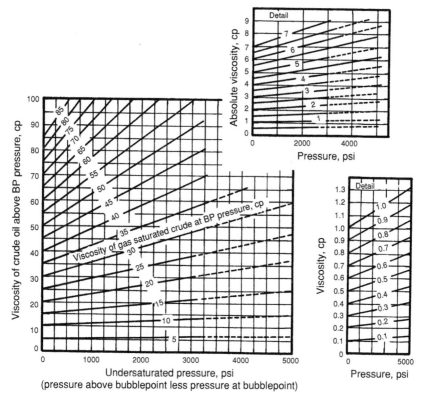

FIGURE 1.11 Increase in oil viscosity with pressure above bubble-point pressure [5].

from the viscosity at the bubble-point pressure, and the difference between reservoir pressure and bubble-point pressure.

Correlations [7] were presented in equation form for the estimation of both dead oil and saturated oil viscosities. These correlations, which are presented in the section on programs for hand-held calculations, neglect the dependence of oil viscosity on composition of the crude. If compositional data are available, other correlations [8–10] for oil viscosity can be used.

Water Viscosity   In 1952, the National Bureau of Standards conducted tests [11] which determined that the absolute viscosity of pure water was 1.0019 cp as compared with the value of 1.005 cp that had been accepted for many years. Effective July 1, 1952, the value of 1.002 cp for the absolute viscosity of water was recommended as the basis for the calibration of viscometers and standard oil samples. Any literature values based on the old standard are in slight error. Water viscosity decreases as temperature is increased as shown in Table 1.1.

Although the predominate effect on water viscosity is temperature, viscosity of water normally increases as salinity increases. Potassium chloride is an exception to this generality. Since most oilfield waters have a high sodium chloride content, the effect of this salt on viscosity of water is given in Table 1.2.

For temperatures of interest in oil reservoirs ($>60°F$), the viscosity of water increases with pressure but the effect is slight. Dissolved gas at reservoir conditions should reduce the viscosity of brines; however, the lack of data and the slight solubility of gas in water suggest that this effect is

TABLE 1.1   Viscosity of Pure Water at Various Temperatures

T, °C	T, °F	Viscosity (cp)
0	32	1.787
10	50	1.307
20	68	1.002
25	77	0.8904
30	86	0.7975
40	104	0.6529
50	122	0.5468
60	140	0.4665
70	158	0.4042
80	176	0.3547
90	194	0.3147
100	212	0.2818

From Reference 12.

TABLE 1.2 Viscosities of Sodium
Chloride Solutions at 68°F

NaCl (wt %)	Viscosity (cp)
0.1	1.004
0.3	1.008
0.5	1.011
1.0	1.020
1.5	1.028
2.0	1.036
3.0	1.052
4.0	1.068
5.0	1.085
10.0	1.193
15.0	1.351
20.0	1.557
25.0	1.902

From Reference 12.

usually ignored. Figure 1.12 is the most widely cited data for the effect of sodium chloride and reservoir temperature on water viscosity [13].

1.1.4 Formation Volume Factors

These factors are used for converting the volume of fluids at the prevailing reservoir conditions of temperature and pressure to standard surface conditions of 14.7 psia and 60°F.

Gas Formation Volume Factor The behavior of gas can be predicted from:

$$pV = znRT \qquad (1.1)$$

where p = absolute pressure
V = volume of gas
T = absolute temperature
n = number of moles of gas
R = gas constant
z = factor to correct for nonideal gas behavior

For conventional field units, p is in psia, V is in ft^3, T is in °R (°F+460), z is dimensionless, n is in lb moles, and R is 10.73 psia ft^3/lb mole °R [14]. The gas formation volume factor, B_g, is the volume of gas in the reservoir occupied by a standard ft^3 of gas at the surface:

$$B_g = \frac{V \, \text{ft}^3}{5.615 \, \text{ft}^3/\text{bbl}} = \frac{znR(T_R + 460)}{5.615} \qquad (1.2)$$

where T_R is the reservoir temperature in °F.

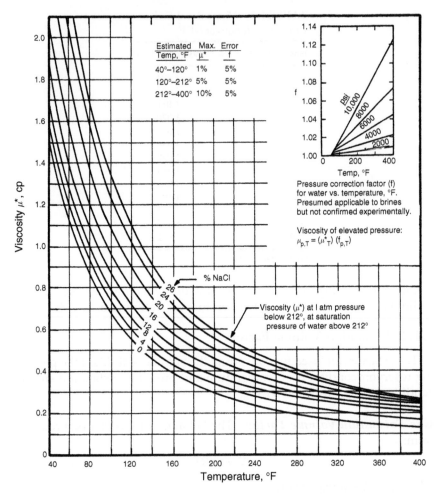

FIGURE 1.12   Water viscosities for various salinities and temperatures [13].

Since one lb mole is equivalent to 379 ft³ at 60°F and 14.7 psia [15]:

$$B_g = \frac{\frac{1}{379} \times 10.73}{5.615} \frac{zT_R}{p} = 0.00504 \frac{zT_R}{p} \tag{1.3}$$

In this expression, B_g will be in reservoir barrels per standard ft³ (RB/scf).

Gas formation volume factor can also be expressed in units of reservoir barrels per stock tank barrel or ft³ of gas at reservoir conditions per ft³ of gas at standard conditions:

$$B_g = 0.02827(460 + T_R)\frac{z}{p} \tag{1.4}$$

Because the gas formation volume factor can be expressed in so many different units (including the reciprocal of B_g), caution should be exercised when B_g is used. In much of the petroleum literature, notably SPE, B_g is expressed in RB/scf. If units of ft^3/scf are given, B_g can be divided by 5.615 or multiplied by 0.1781 to get RB/scf.

Gas formation volume factors can be estimated by determining the gas deviation factor or compressibility factor, z, at reservoir pressure, p, and temperature T_R from the correlations of Standing and Katz [16] (Figure 1.13). To obtain the z factor, reduced pressure, p_r, and reduced temperature, T_r, are calculated:

$$p_r = \frac{p}{p_c} \tag{1.5}$$

where p_c is the critical pressure and

$$T_r = \frac{T}{T_c} \tag{1.6}$$

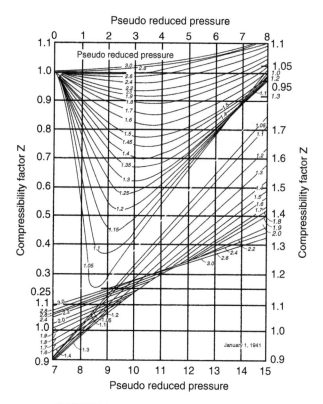

FIGURE 1.13 Compressibility factors [16].

where T_c is the critical temperature. The critical pressure and temperature represent conditions above which the liquid and vapor phase are indistinguishable.

Compressibility factor and gas formation volume factor can be more conveniently estimated by the use of programs available for hand-held calculators. These programs will be subsequently discussed.

Oil Formation Volume Factor   The volume of hydrocarbon liquids produced and measured at surface conditions will be less than the volume at reservoir conditions. The primary cause is the evolution of gas from the liquids as pressure is decreased from the reservoir to the surface. When there is a substantial amount of dissolved gas, a large decrease in liquid volume occurs. Other factors that influence the volume of liquids include changes in temperature (a decrease in temperature will cause the liquid to shrink) and pressure (a decrease in pressure will cause some liquids to expand). All of these factors are included in the oil formation volume factor, B_o, which is the volume of oil in reservoir barrels, at the prevailing reservoir conditions of pressure and temperature, occupied by a stock tank barrel of oil at standard conditions. The withdrawal of reservoir fluids can be related to surface production volumes by obtaining laboratory PVT data with reservoir fluids. Such data include B_g (the gas formation volume factor), B_o (the oil formation volume factor), and R_s (the solution gas-oil ratio which is the volume of gas in standard ft^3 that will dissolve in one stock tank barrel of oil at reservoir conditions).

The formation volume factor is used to express the changes in liquid volume accompanied by changes in pressure. Changes in formation volume factor with pressure for an undersaturated crude are displayed in Figure 1.14 [17]. As the initial reservoir pressure decreases, the all-liquid system expands and the formation volume factor increases until the bubble-point pressure is reached. As pressure decreases below the bubblepoint, gas comes out of solutions, the volume of oil is reduced, and B_o decreases. For a saturated crude, the trend would be similar to that observed to the left of bubble-point pressure in Figure 1.14.

When two phases exist, the total formation volume factor or 2-phase formation volume factor is [17]:

$$B_t = B_o + B_g(R_{si} - R_s) \tag{1.7}$$

which includes the liquid volume, B_o, plus the gas volume times the difference in initial solution gas-oil ratio, R_{si}, and the solution gas-oil ratio at the specific pressure, R_s. At pressures above the bubblepoint, R_{si} equals R_s, and the single-phase and 2-phase formation volume factors are identical. At pressures below the bubblepoint, the 2-phase factor increases as pressure is decreased because of the gas coming out of solution and the expansion of the gas evolved.

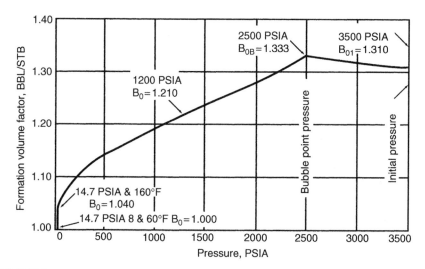

FIGURE 1.14 Formation volume factor of the Big Sandy Field reservoir oil, by flash liberation at reservoir temperature of 160°F [17].

For a system above the bubblepoint pressure, B_o is lower than the formation volume factor at saturation pressure because of contraction of the oil at higher pressure. The customary procedure is to adjust the oil formation volume factor at bubble-point pressure and reservoir temperature by a factor that accounts for the isothermal coefficient of compressibility such as [18]:

$$B_o = B_{ob} \exp[-c_o(p - p_b)] \tag{1.8}$$

where B_{ob} is the oil formation volume factor at bubblepoint conditions, p_b is the bubble-point pressure in psi, and c_o is oil compressibility in psi^{-1}.

The basic PVT properties (B_o, R_s, and B_g) of crude oil are determined in the laboratory with a high-pressure PVT cell. When the pressure of a sample of crude oil is reduced, the quantity of gas evolved depends on the conditions of liberation. In the flash liberation process, the gas evolved during any pressure reduction remains in contact with the oil. In the differential liberation process, the gas evolved during any pressure reduction is continuously removed from contact with the oil. As a result, the flash liberation is a constant-composition, variable-volume process and the differential liberation is a variable-composition, constant-volume process. For heavy crudes (low volatility, low API gravity oils) with dissolved gases consisting primarily of methane and ethane, both liberation processes yield similar quantities and compositions of evolved gas as well as similar resulting oil volumes. However, for lighter, highly volatile crude oils containing a relatively high proportion of intermediate hydrocarbons (such as propane,

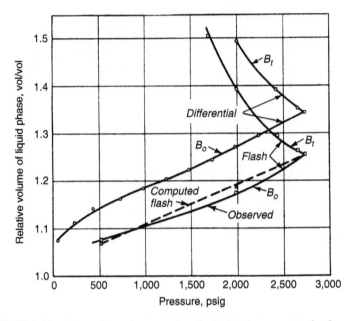

FIGURE 1.15   Comparison of measured and calculated composite oil volume [19].

butane, and pentane), the method of gas liberation can have an effect on the PVT properties that are obtained. An example of differences in formation volumes with flash and differential liberation processes can be seen in Figure 1.15 [19]. Actual reservoir conditions may be somewhere between these extremes because the mobility of the liberated gas is greater than the oil, the gas is produced at a higher rate, and the oil in the reservoir is in contact with all of the initial solution gas for only a brief period [20]. Since volatile oil situations are uncommon [20], many engineers feel the differential liberation process typifies most reservoir conditions [19]. For reservoir fluids at the bubblepoint when a well is put on production, the gas evolved from the oil as the pressure declines does not flow to the well until the critical gas saturation is exceeded. Since the greatest pressure drop occurs near the wellbore, the critical gas saturation occurs first near the well, especially if the pressure drop is large. In general, differential liberation data are applicable if the reservoir pressure falls considerably below the bubblepoint pressure and the critical gas saturation is exceeded in the majority of the drainage area, as indicated by producing gas-oil ratios considerably in excess of the initial solution gas-oil ratio [17]. Flash liberation data may be applicable to reservoirs where there is only a moderate pressure decline below the bubblepoint, as indicated by producing gas-oil ratios not much higher than the initial solution gas-oil ratio, since the liberated gas stays in the reservoir in contact with the remaining oil [17].

Several correlations are available for estimating formation volume factors. Single-phase formation volume factors can be estimated from solution gas, gravity of solution gas, API gravity of the stock tank oil, and reservoir temperature by using the correlations of Standing [1, 2]. Figure 1.3 provides Standing's empirical correlation of bubble-point oil formation volume factor as a function of the variables mentioned. Total formation volume factors of both solution gas and gas-condensate systems can be obtained from Standing's correlations given in Figure 1.4.

Empirical equations have been developed [21] from Standing's graphical data. These equations provide the oil formation volume factor and the solution gas-oil ratio as functions of reservoir pressure [21]:

$$B_o = a\,p^{1.17} + b \tag{1.9}$$

where is a constant that depends on temperature, oil API gravity and gas gravity and b is a constant that depends on temperature. Values of both constants are given in Table 1.3, other values can be interpolated.

TABLE 1.3 Values of Constants for Equation 1.9

		Values of $a \times 10^5$					
Oil gravity		$T = 120°F$			$T = 140°F$		
°API	Gas gravity:	0.7	0.8	0.9	0.7	0.8	0.9
26		2.09	2.55	3.10	2.03	2.58	3.13
30		2.44	2.98	3.61	2.38	3.01	3.64
34		2.85	3.48	4.21	2.78	3.51	4.24
38		3.33	4.07	4.90	3.26	4.10	4.93
42		3.89	4.75	5.71	3.82	4.78	5.74
Oil gravity		$T = 160°F$			$T = 180°F$		
°API	Gas gravity:	0.7	0.8	0.9	0.7	0.8	0.9
26		2.02	2.47	3.02	1.95	2.38	2.91
30		2.33	2.85	3.48	2.27	2.78	3.39
34		2.69	3.29	4.01	2.65	3.24	3.96
38		3.10	3.80	4.62	3.09	3.79	4.61
42		3.58	4.38	5.33	3.60	4.42	5.38

Values of b	
T, °F	b
120	1.024
140	1.032
160	1.040
180	1.048

From Reference 21.

Solution gas-oil ratio can be estimated from:

$$R_s = y \, p^{1.17} \qquad (1.10)$$

where y is a constant that depends on temperature, gas gravity, and oil gravity. Values of y are provided in Table 1.4.

Water Formation Volume Factor   The factors discussed that affected B_o also affect the water formation volume factor, B_w. However, gas is only slightly soluble in water so evolution of gas from water has a negligible effect on B_w. Expansion and contraction of water due to reduction of pressure and temperature are slight and offsetting. Hence, B_w is seldom greater than 1.06 [18] and is usually near unity (see Table 1.5).

Several correlations for B_w are available, including the effect of gas saturation in pure water and the effect of salinity [23], and the effect of natural gas on B_w as a function of pressure and temperature [24]. However, since B_w is not greatly affected by these variables, only a simplified correction is presented [18]:

$$B_w = (1 + \Delta V_{wp})(1 + \Delta V_{wT}) \qquad (1.11)$$

where ΔV_{wp} and ΔV_{wT} are the volume changes caused by reduction in pressure and temperature, respectively. Values of these corrections are given in Figures 1.16 and 1.17.

TABLE 1.4   Values of Constants for Equation 1.10

Oil gravity		$T = 120°F$			$T = 140°F$		
°API	Gas gravity:	0.7	0.8	0.9	0.7	0.8	0.9
26		0.0494	0.0577	0.0645	0.0481	0.0563	0.0632
30		0.0568	0.0660	0.0737	0.0550	0.0635	0.0721
34		0.0654	0.0755	0.0842	0.0630	0.0736	0.0823
38		0.0752	0.0864	0.0962	0.0720	0.0841	0.0939
42		0.0865	0.0989	0.1099	0.0824	0.0961	0.1071
Oil gravity		$T = 160°F$			$T = 180°F$		
°API	Gas gravity:	0.7	0.8	0.9	0.7	0.8	0.9
26		0.0453	0.0519	0.0591	0.0426	0.0481	0.0543
30		0.0522	0.0597	0.0677	0.0492	0.0557	0.0629
34		0.0601	0.0686	0.0775	0.0567	0.0645	0.0728
38		0.0692	0.0788	0.0887	0.0654	0.0747	0.0842
42		0.0797	0.0906	0.1016	0.0755	0.0865	0.0975

From Reference 21.

TABLE 1.5 Formation Volumes of Water

Pressure psia	Formation volumes, bbl/bbl			
	100°F	150°F	200°F	250°F
		Pure water		
5,000	0.9910	1.0039	1.0210	1.0418
4,000	0.9938	1.0067	1.0240	1.0452
3,000	0.9966	1.0095	1.0271	1.0487
2,000	0.9995	1.0125	1.0304	1.0523
1,000	1.0025	1.0153	1.0335	1.0560
Vapor pressure of water	1.0056	1.0187	1.0370	1.0598

Saturation pressure psia	Natural gas and water			
5,000	0.9989	1.0126	1.0301	1.0522
4,000	1.0003	1.0140	1.0316	1.0537
3,000	1.0017	1.0154	1.0330	1.0552
2,000	1.0031	1.0168	1.0345	1.0568
1,000	1.0045	1.0183	1.0361	1.0584

From Reference 22.

1.1.5 Fluid Compressibilities

Gas Compressibility The compressibility of a gas, which is the coefficient of expansion at constant temperature, should not be confused with the compressibility factor, z, which refers to the deviation from ideal gas behavior. From the basic gas equation (see Equation 1.2), Muskat [25] provided an expression for the coefficient of isothermal compressibility:

$$c_g = \frac{1}{p} - \frac{1}{z}\frac{dz}{dp} \qquad (1.12)$$

For perfect gases ($z=1$ and $dz/dp=0$), c_g is inversely proportional to pressure. For example, an ideal gas at 1,000 psia has a compressibility of $1/1,000$ or $1,000 \times 10^{-6}$ psi^{-1}. However, natural hydrocarbon gases are not ideal gases and the compressibility factor, z, is a function of pressure as seen in Figure 1.18 [17]. At low pressures, z decreases as pressure increases and dz/dp is negative; thus, c_g is higher than that of an ideal gas. At high pressures, dz/dp is positive since z increases, and c_g is less than that of a perfect gas.

Compared to other fluids or to reservoir rock, the compressibility of natural gas is large; c_g ranges from about $1,000 \times 10^{-6}$ psi^{-1} at 1,000 psi to about 100×10^{-6} psi^{-1} at 5,000 psi [27]. Compressibility of natural gases can be obtained from laboratory PVT data or estimated from the correlations given by Trube [27] (see Figures 1.19a and 1.19b). Trube defined the pseudo-reduced compressibility of a gas, c_{pr}, as a function of

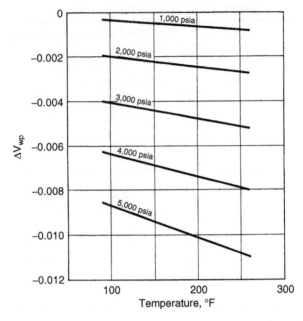

FIGURE 1.16   Change in water volume due to pressure reduction [18].

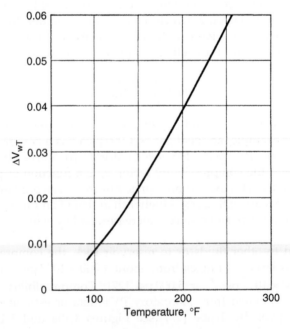

FIGURE 1.17   Change in water volume due to temperature reduction [18].

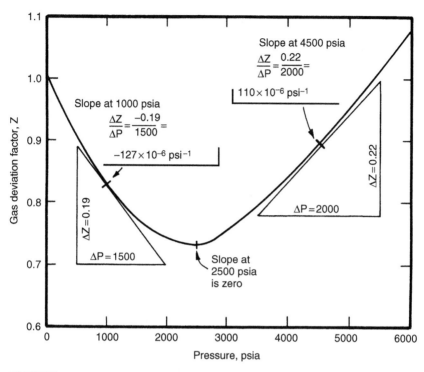

FIGURE 1.18 Gas compressibility from the gas deviation factor versus pressure [17].

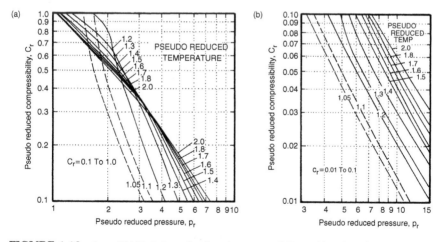

FIGURE 1.19 (a and b) Variation of reduced compressibility with reduced pressures for various fixed values of reduced temperature [27].

pseudo-reduced temperature and pressure, T_{pr} and p_{pr}, respectively [27]:

$$c_g = \frac{1}{p_{pc}} \left[\frac{1}{p_{pr}} - \frac{1}{z} \left(\frac{dz}{dp_{pr}} \right) T_{pr} \right] \qquad (1.13)$$

where p_{pc} is the pseudo-critical pressure (reduced and critical pressures have been defined earlier). Gas compressibility is computed for the pseudo-reduced compressibility from the appropriate figure:

$$c_g = \frac{c_{pr}}{p_{pr}} \qquad (1.14)$$

Pseudo-critical pressures and temperatures can be calculated from the mole fraction of each component present in hydrocarbon gas mixture or estimated from Figure 1.20 [3].

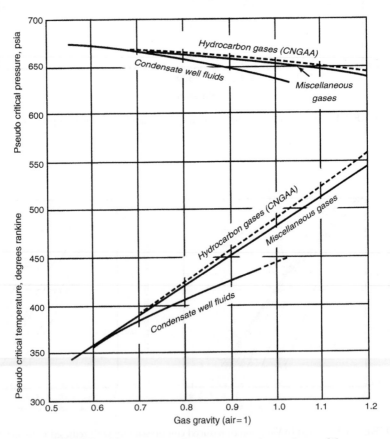

FIGURE 1.20   Pseudo-critical properties of natural gases [3].

Oil Compressibility The compressibility of oil, c_o, can be obtained in the laboratory from PVT data. In the absence of laboratory data, Trube's correlation [28] for compressibility of an undersaturated oil in Figure 1.21 can be used in a similar fashion as previously discussed for c_g. Pseudo-critical temperature and pressure can be estimated from Figure 1.22 or 1.23.

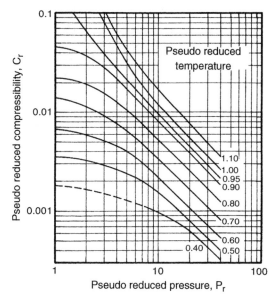

FIGURE 1.21 Variation of pseudo-reduced compressibility with pseudo-reduced pressures for various fixed values of pseudo-reduced temperature [28].

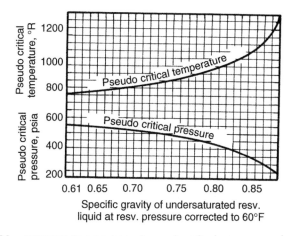

FIGURE 1.22 Approximate variation of pseudo-critical pressure and pseudo-critical temperature with specific gravity of liquid corrected to 60°F [28].

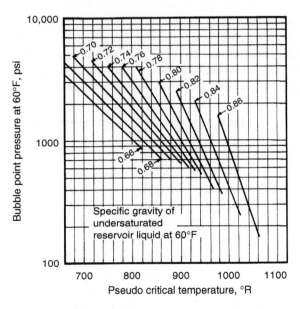

FIGURE 1.23 Variation of pseudo-critical temperature with specific gravity and bubble point of liquid corrected to 60°F [28].

With the pseudo-reduced compressibility from Figure 1.21, oil compressibility can be estimated:

$$c_o = \frac{c_{pr}}{p_{pc}} \qquad (1.15)$$

For conditions below the bubblepoint, dissolved gas must be taken into account. In the absence of laboratory data, the changes in R_s and B_o with changes in pressure can be approximated from Figures 1.24 and 1.25, which were developed by Ramey [26] from Standing's [1] data:

$$\left(\frac{\partial B_o}{\partial p}\right)_T = \left(\frac{\partial R_s}{\partial p}\right)_T \left(\frac{\partial B_o}{\partial R_s}\right)_T \qquad (1.16)$$

B_o can be estimated from Figure 1.3, and gravities of both oil and gas must be known. Oil compressibility is often on the order of $10 \times 10^{-6}\,\text{psi}^{-1}$.

There are several types of petroleum-engineering software equipped with the so-called "PVT data" generator code. Engineers only need fill in the known fluid properties or pressure-temperature data. By choosing the suitable correlation, the software will provide all pertinent PVT data.

Water Compressibility Although the best approach is to obtain water compressibilities from laboratory PVT tests, this is seldom done and the

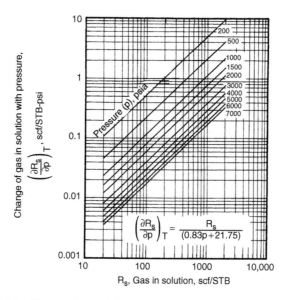

FIGURE 1.24 Change of gas solubility in oil with pressure vs. gas in solution [26].

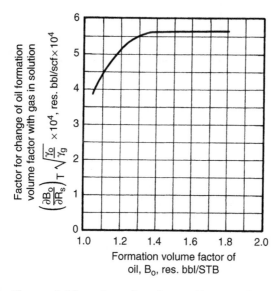

FIGURE 1.25 Change of oil formation volume factor with gas in solution vs. oil formation volume factor [26].

use of correlations [22] such as are given in Figures 1.26 and 1.27 is often required. The compressibility of nongas-saturated water ranges from 2×10^{-6} psi^{-1} to 4×10^{-6} psi^{-1} and a value of 3×10^{-6} psi^{-1} is frequently used [13]. The compressibility of water with dissolved gas ranges from 15×10^{-6} psi^{-1} at 1,000 psi to 5×10^{-6} psi^{-1} at 5,000 psi [26].

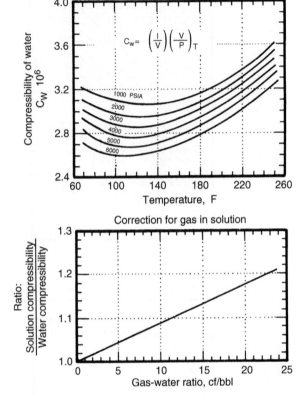

FIGURE 1.26   Effect of dissolved gas on the compressibility of water [22].

1.1.6 Estimation of Fluid Properties with Programmable Calculators and Personal Computers

With the widespread use of hand-held programmable calculators and desk-top personal computers, engineers are no longer faced with estimating fluid properties from charts and graphs. Much of the data in the literature have been transmitted to empirical equations suitable for use with programmable calculators or personal computers. In some cases, improved empirical data have now been presented. This section provides a number of the expressions available for computer use and also provides references for books devoted to programs for hand-held calculators. References to some of the software available for personal computers will be given.

Data from Figure 1.12 have been used to develop a correlation for water viscosity that can be used on programmable calculators [29]:

$$\mu_w = \left(A + \frac{B}{T}\right) f_{pt} \tag{1.17}$$

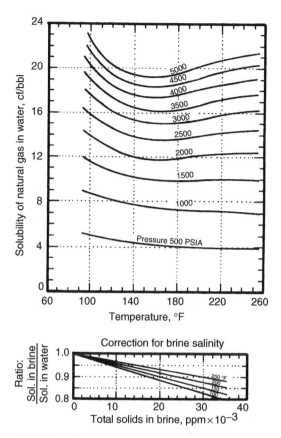

FIGURE 1.27 Solubility of natural gas in water [22].

where $A = -0.04518 + 0.009313\,(\%\ NaCl) - 0.000393\,(\%\ NaCl)^2$
$\quad B = 70.634 + 0.09576\,(\%\ NaCl)^2$
$\quad T = \text{temperature, }°F$
$\quad f_{pt} = 1 + [3.5E - 12p^2(T - 40)]$

The correlation should only be used as an estimate and applies for pressures less than 10,000 psi, salinity less than 26% NaCl, and in a temperature range of 60° to 400°F.

In 1977, Standing's classic work [2] was reprinted [30] by the Society of Petroleum Engineers and an appendix was added by Standing that provides equations for several of the charts in the original work. Most of the equations were developed by simple curve fitting procedures. Some equations were based on computer solutions by other individuals; details of this will not be presented here and the reader is referred to Appendix 2 of Reference 30. Gas viscosity can be estimated from the correlations of Carr, Kobayashi, and Burrows [31] (the basis of Figures 1.6 and 1.7); first the atmospheric value of gas gravity at reservoir temperature, estimated from gravity and nonhydrocarbon content:

$$\mu_1 = (\mu_1 \text{ uncorrected}) + (N_2 \text{ correction})$$
$$+ (CO_2 \text{ correction}) + (H_2S \text{ correction}) \tag{1.18}$$

where $(\mu_1 \text{ uncorrected}) = \left[1.709\left(10^{-5}\right) - 2.062\left(10^{-6}\right)\gamma_g\right]$
$$\times T + 8.188\left(10^{-3}\right) - 6.15\left(10^{-3}\right)\log\gamma_g$$
$$(N_2 \text{ correction}) = y_{N_2}\left[8.48\left(10^{-3}\right)\log\gamma_g + 9.59\left(10^{-3}\right)\right]$$
$$(CO_2 \text{ correction}) = y_{CO_2}\left[9.08\left(10^{-3}\right)\log\gamma_g + 6.24\left(10^{-3}\right)\right]$$
$$(H_2S \text{ correction}) = y_{H_2S}\left[8.49\left(10^{-3}\right)\log\gamma_g + 3.73\left(10^{-3}\right)\right]$$

is adjusted to reservoir conditions by a factor based on reduced temperature and pressure:

$$\ln\left[\left(\frac{\mu_g}{\mu_i}\right)(T_{pr})\right] = a_0 + a_1 p_{pr} + a_2 p_{pr}^2 + a_3 p_{pr}^3$$
$$+ T_{pr}\left(a_4 + a_5 p_{pr} + a_6 p_{pr}^2 + a_7 p_{pr}^3\right)$$
$$+ T_{pr}^2\left(a_8 + a_9 p_{pr} + a_{10} p_{pr}^2 + a_{11} p_{pr}^3\right)$$
$$+ T_{pr}^3\left(a_{12} + a_{13} p_{pr} + a_{14} p_{pr}^2 + a_{15} p_{pr}^3\right) \tag{1.19}$$

where $a_0 = -2.462\,118\,20E - 00$ $a_8 = -7.933\,856\,84E - 01$
      $a_1 = 2.970\,547\,14E - 00$ $a_9 = 1.396\,433\,06E - 00$
      $a_2 = -2.862\,640\,54E - 01$ $a_{10} = -1.491\,449\,25E - 01$
      $a_3 = 8.054\,205\,22E - 03$ $a_{11} = 4.410\,155\,12E - 03$
      $a_4 = 2.808\,609\,49E - 00$ $a_{12} = 8.393\,871\,78E - 02$
      $a_5 = -3.498\,033\,05E - 00$ $a_{13} = -1.864\,088\,48E - 01$
      $a_6 = 3.603\,703\,20E - 01$ $a_{14} = 2.033\,678\,81E - 02$
      $a_7 = -1.044\,324\,13E - 02$ $a_{15} = -6.095\,792\,63E - 04$
                                         $(= 0.000\,609\,57\,9263)$

A reasonable fit to Beal's correlation (Figure 1.8 of Reference 5) of gas-free or dead oil viscosity (which is not very precise) is given by Standing:

$$\mu_{oD} = \left[0.32 + \frac{1.8(10^7)}{°API^{4.53}}\right]\left(\frac{360}{T + 200}\right)^a \tag{1.20}$$

where $a = \text{antilog}\left(0.43 + \frac{8.33}{°API}\right)$

The dead oil viscosity can then be adjusted for dissolved gas with the correlation by Chew and Connally [6] (Figure 1.10) for saturated oil:

$$\mu_{ob} = A(\mu_{oD})^b \tag{1.21}$$

where A and b are functions of solution gas-oil ratio.

Fit of A and b values given in Table 1.3 of Reference 6 are

$$A = \text{antilog}\{R_s[2.2(10^{-7})R_s - 7.4(10^{-4})]\}$$

$$b = \frac{0.68}{10^{8.62}(10^{-5})R_s} + \frac{0.25}{10^{1.1}(10^{-3})R_s} + \frac{0.062}{10^{3.74}(10^{-3})R_s}$$

The equation for compressibility factors of natural gases (Figure 1.13) is:

$$Z = A + (1-A)/e^B + Cp_{pr}^D \qquad (1.22)$$

where $A = 1.39(T_{pr} - 0.92)^{0.5} - 0.36T_{pr} - 0.101$

$$B = (0.62 - 0.23T_{pr})p_{pr} + \left[\frac{0.066}{(T_{pr} - 0.86)} - 0.037\right]p_{pr}^2 + \frac{0.32}{10^9(T_{pr}^{-1})}p_{pr}^6$$

$$C = (0.132 - 0.32\log T_{pr})$$

$$D = \text{antilog}(0.3106 - 0.49T_{pr} + 0.1824T_{pr}^2)$$

T_{pr} is the dimensionless pseudo-reduced temperature, and p_{pr} is the dimensionless pseudo-reduced pressure. The relationship between formation volume factor of bubble-point liquids and gas-oil ratio, dissolved gas gravity, API oil gravity, and temperature (Figure 1.3) is [30]:

$$B_{ob} = 0.9759 + 12(10^{-5})(CN)B_{ob}1.2 \qquad (1.23)$$

where

$$(CN)_{B_{ob}} = R_s\left(\frac{\gamma_g}{\gamma_o}\right)^{0.5} + 1.25T$$

where $(CN)B_{ob} =$ bubble-point formation volume factor correlating number
$R_s =$ solution gas-oil ratio in ft^3/bbl
$\gamma_g =$ gas gravity (air $= 1$)
$\gamma_o =$ oil specific gravity (water $= 1$)
$T =$ temperature in °F

The correlation of bubble-point pressure with dissolved gas-oil ratio, dissolved gas gravity, API oil gravity, and temperature (Figure 1.2) is [30]:

$$P_b = 18.2[(CN)_{pb} - 1.4] \qquad (1.24)$$

where

$$(CN)_{pb} = \left(\frac{R_s}{\gamma_g}\right)^{0.83}\left[10^{(0.00091T - 0.125°API)}\right]$$

where $(CN)_{pb}$ is the bubble-point pressure correlating number and the other terms have been previously defined.

Standing also presents equation for; density correction for compressibility of liquids; density correction for thermal expansion of liquids; apparent liquid densities of natural gases; effect of condensate volume on the ratio of surface gas gravity to well fluid gravity; pseudo-critical constants of gases and condensate fluids; pseudo-liquid density of systems containing methane and ethane; and pseudo-critical temperatures and pressures for heptane and heavier.

Beggs and Robinson [7] collected PVT data and presented a better estimate of the dead oil viscosity as a function of temperature and oil specific gravity:

$$\mu_{OD} = 10^X - 1 \tag{1.25}$$

where μ_{OD} = viscosity in cp of the gas-free oil at temperature, T, and $X = yT^{-1.163}$, $y = 10^Z$, and $Z = 3.0324 - 0.02023\,\gamma_o$, with T in °F and the oil gravity γ in °API. An expression was also given for the saturated oil viscosity, μ, or live oil below the bubblepoint which results from a linear relationship between $\log \mu_{OD}$ and $\log \mu$ for a given value of dissolved gas, R_s:

$$\mu = A\mu_{OD}^B \tag{1.26}$$

where $A = 10.715\,(R_s + 100)^{-0.515}$
$\quad\quad B = 5.44(R_s + 150)^{-0.338}$
$\quad\quad R_s = \text{scf/STB}$

In the first book specifically for hand-held calculators, Hollo and Fifadara [32] presented programs for estimating gas deviation factor (based on data of Standing and Katz):

$$Z = 1 + (A_1 + A_2/T_R + A_3/T_R^3)\rho R$$
$$+ (A_4 + A_5/T_R)\rho R^2 + A_6\rho R^2/T_R^3 \tag{1.27}$$

where $\rho R = 0.27\,P_R/ZT_R$ $\quad T_R = T/T_C$ $\quad P_R = P/P_C$
$\quad\quad A_1 = 0.31506$ $\quad\quad A_2 = -1.0467$ $\quad A_3 = -0.5783$
$\quad\quad A_4 = 0.5353$ $\quad\quad A_5 = -0.6123$ $\quad A_6 = 0.6815$

A program was also presented to calculate the single-phase formation volume factor using the correlation developed by Standing:

$$B_o = 0.972 + \frac{1.47}{10^4}\left[(R_s)\left(\frac{\gamma_g}{\gamma_o}\right)^{0.5} + 1.25T\right]^{1.75} \tag{1.28}$$

where γ_g = solution gas specific gravity
$\quad\quad \gamma_o$ = stock tank oil specific gravity $(141.5/131.5 + °API)$
$\quad\quad T$ = temperature,°F

R_s = solution gas-oil ratio, scf/STB
B_o = single-phase formation volume factor, RB/STB
°API = stock tank oil gravity, °API

In 1980 Vazquez and Beggs [33] published improved empirical correlations for some of the commonly required crude oil PVT properties. Their study utilized a much larger database than was used in previous work so the results are applicable to a wider range of oil properties. The empirical correlations, presented as a function of gas specific gravity, oil API gravity, reservoir temperature, and pressure, are particularly convenient to use with hand-held calculators. Gas gravity was found to be a strong correlating parameter. Since gas gravity depends on gas-oil separation conditions, Vazquez and Beggs chose 100 psig as a reference pressure, which resulted in a minimum oil shrinkage for the separator tests available. Thus, gas gravity must first be corrected to the value that would result from separation at 100 psig:

$$\gamma_{gs} = \gamma_{gp}[1 + 5.912 \times 10^{-5}(\gamma_o)(T)\log(p/114.7)] \tag{1.29}$$

where γ_{gs} = gas gravity (air = 1) that would result from separator
conditions of 100 psig
γ_{gp} = gas gravity obtained at separator conditions of p and T
p = actual separator pressure, psia
T = actual separator temperature, °F
γ_o = oil gravity, °API

For both dissolved gas and oil formation volume factor, improved correlations were obtained when the measured data were divided into two groups, with the division made at an oil gravity of 30°API. The expression for dissolved gas was presented:

$$R_s = C_1\gamma_{gs}p^{C_2}\exp\{C_3[\gamma_o/(T+460)]\} \tag{1.30}$$

Values for the coefficients are as follows.

Coefficient	$\gamma_o \leq 30$	$\gamma_o > 30$
C_1	0.0362	0.0178
C_2	1.0937	1.1870
C_3	25.7240	23.9310

For saturated oils (reservoir pressure less than bubblepoint), oil formation volume factor was expressed as:

$$B_o = 1 + C_1R_s + C_2(T-60)(\gamma_o/\gamma_{gs}) + C_3R_s(T-60)(\gamma_o/\gamma_{gs}) \tag{1.31}$$

The values for the coefficients depend on oil gravity and are given by the following:

Coefficient	$\gamma_o \leq 30$	$\gamma_o > 30$
C_1	4.677×10^{-4}	4.670×10^{-4}
C_2	1.751×10^{-5}	1.100×10^{-5}
C_3	-1.811×10^{-8}	1.337×10^{-9}

Since the oil formation volume of an undersaturated crude depends on the isothermal compressibility of the liquid, the oil formation volume as pressure is increased above bubble-point pressure was calculated from:

$$B_o = B_{ob} \exp[c_o(p - p_b)] \tag{1.32}$$

The correlation for compressibility of oil was given as:

$$c_o = (a_1 + a_2 R_s + a_3 T + a_4 \gamma_{gs} + a_5 \gamma_o)/a_6 p \tag{1.33}$$

where $a_1 = -1{,}433.0$
$a_2 = 5.0$
$a_3 = 17.2$
$a_4 = -1{,}180.0$
$a_5 = 12.61$
$a_6 = 105$

Vazquez and Beggs also presented an equation for viscosity of undersaturated crude oils that used the correlations of Beggs and Robinson:

$$\mu_o = \mu_{ob}(p/p_b)^m \tag{1.34}$$

where $m = C_1 p^{C_2} \exp(C_3 + C_4 p)$
$C_1 = 2.6$
$C_2 = 1.187$
$C_3 = -11.513$
$C_4 = -8.98 \times 10^{-5}$

The improved correlations of Vazquez and Beggs were incorporated by Meehan in the development of programs for hand-held calculators. These programs were presented in a series of articles in the *Oil and Gas Journal* [34, 35]. Reference 34 contains the programs for calculating gas gravity, dissolved gas-oil ratio, oil formation volume factor, and oil compressibility. Reference 35 contains the program for calculating oil viscosity.

See References 36–40 for a list of books devoted to the use of programs for handheld calculators and personal computers.

1.2 PROPERTIES OF FLUID-CONTAINING ROCKS

1.2.1 Porosity

The porosity, ϕ, is equal to the void volume of the rock divided by the bulk volume and is expressed as a percent or fraction of the total bulk volume of the rock. Oil-bearing sandstones have porosities which often range from 15% to 30%. Porosities of limestones and dolomites are usually lower.

Differentiation must be made between absolute and effective porosity. Absolute porosity is defined as the ratio of the total pore volume of the rock to the total bulk volume of the rock whereas effective porosity is defined as the ratio of the interconnected pore volume of the rock to the total bulk volume of the rock.

Factors affecting porosity are compactness, character and amount of cementation, shape and arrangement of grains, and uniformity of grain size or distribution.

In problems involving porosity calculations it is convenient to remember that a porosity of one percent is equivalent to the presence of 77.6 barrels of pore space in a total volume of one acre-foot of sand.

1.2.2 Pore Volume

The pore volume of a reservoir is the volume of the void space, that is, the porosity fraction times the bulk volume. In conventional units, the pore volume, V_p, in reservoir barrels is:

$$V_p = 7758 \qquad V_b\phi = 7,758\,A\,h\,\phi \qquad (1.35)$$

where V_b is the bulk volume in ac-ft. A is the area in ft^2, h is the reservoir thickness in ft. and ϕ is the porosity expressed as a fraction.

1.2.3 Permeability

The permeability of a rock is a measure of the ease with which fluids flow through the rock. It is denoted by the symbol k and commonly expressed in units of darcies. Typical sandstones in the United States have permeabilities ranging from 0.001 to a darcy or more, and for convenience the practical unit of permeability is the millidarcy which equals 0.001 darcy. Some other useful conversion factors are given in Table 1.6.

1.2.4 Absolute Permeability

If a porous system is completely saturated with a single fluid, the permeability is a rock property and not a property of the flowing fluid (with the exception of gases at low pressure). This permeability at 100% saturation of a single fluid is termed the absolute permeability.

TABLE 1.6   Permeability Conversion Factors

---

1 darcy $= 1,000$ millidarcies; 1 millidarcy $= 0.001$ darcy

$\qquad = 0.9869233 \, \mu m^2 \, (1 \, md \cong 10^{-3} \, \mu m^2)$

$$k = \frac{q\mu}{(A)(\Delta p/L)}$$

$$1 \, darcy = \frac{(cc/sec)(cp)}{(cm^2)(atm)/cm}$$

$$= 9.869 \times 10^{-9} \, cm^2$$

$$= 1.062 \times 10^{-11} \, ft^2$$

$$= 1.127 \frac{[bbl/(day)](cp)}{(ft^2)(psi)/ft}$$

$$= 9.869 \times 10^{-7} \frac{(cc/sec)(cp)}{cm^2[dyne/(cm^2)(cm)]}$$

$$= 7.324 \times 10^{-5} \frac{[ft^3/(sec)](cp)}{(ft^2)(psi)/ft}$$

$$= 9.679 \times 10^{-4} \frac{[ft^3/(sec)](cp)}{(cm^2)(cm \; water)/cm}$$

$$= 1.424 \times 10^{-2} \frac{[gal/(min)](cp)}{(ft^2)(ft \; water)/ft}$$

From Reference 19.

1.2.5 Darcy Equation

The Darcy equation relates the apparent velocity, v, of a homogeneous fluid in a porous medium to the fluid viscosity, μ, and the pressure gradient, $\Delta p/L$:

$$v = -\frac{k\Delta p}{\mu L} \tag{1.36}$$

This equation states that the fluid velocity is proportional to the pressure gradient and inversely proportional to fluid viscosity. The negative sign indicates that pressure decreases in the L direction when the flow is taken to be positive. The flow rate, q, is understood to be positive during production and negative during injection. As shown in Table 1.6, the Darcy equation can be written as:

$$k = \frac{q\mu L}{A\Delta p} \tag{1.37}$$

Linear Flow   In the Darcy equation for linear displacement:

$$q = \frac{kA(\Delta p)}{\mu L} \tag{1.38}$$

where q = fluid flow rate, cm^3/sec
    A = cross-sectional area of rock perpendicular to flow, cm^2
    p = pressure difference (in atm) across the distance L parallel to
        flow direction, cm
    μ = viscosity of fluid, cp

A rock has permeability of one darcy if it permits the flow of one cc per second of a one-phase fluid having viscosity of one centipoise under the pressure gradient of one atmosphere per gradient of one atmosphere per centimeter. For liquid flow in a linear porous system, the flow rate q in barrels per day is [20]:

$$q = \frac{1.127\,kA(p_c - p_w)}{\mu L} \qquad (1.39)$$

where k is in darcies, A is in ft^2, μ is in cp, L is in ft, and pressures are in psia. For laminar gas flow in a linear system [20]:

$$q = \frac{0.112\,kA(p_c^2 - p_w^2)}{\mu zTL} \qquad (1.40)$$

where q is in Mscf/D, T is in °R(°F+460), z is the dimensionless compressibility factor, and the other terms are in units consistent with Equation 1.39.

Radial Flow   Production from or injection into a reservoir can be viewed as flow for a cylindrical region around the wellbore. For the steady-state radial flow of an incompressible fluid [19]:

$$q = \frac{2\pi kh(p_e - p_w)}{\mu \ln(r_e/r_w)} \qquad (1.41)$$

where q = volume rate of flow, cc/sec
    k = permeability, darcies
    h = thickness, cm
    μ = viscosity, cp
    p_e = pressure at external boundary, atm
    p_w = pressure at internal boundary, atm
    r_e = radius to external boundary, cm
    r_w = radius to internal boundary, cm
    $\ln$ = natural logarithm, base e

For the flow rate, q, in the barrels per day of a liquid [19]:

$$q = \frac{7.08\,kh(p_e - p_w)}{\mu \ln(r_e/r_w)} \qquad (1.42)$$

where k is in darcies, h is in ft., pressures are in psia, μ is in cp, and the radii are in consistent units, usually feet. For the laminar flow of a gas in MscfD [20]:

$$q = \frac{0.703\,kh(p_e^2 - p_w^2)}{\mu z T \ln(r_e/r_w)} \tag{1.43}$$

where T is in °R, z is the dimensionless compressibility factor, and the other terms are as defined in Equation 1.42.

1.2.6 Capacity

Flow capacity is the product of permeability and reservoir thickness expressed in md ft. Since the rate of flow is proportional to capacity, a 10-ft thick formation with a permeability of 100 md should have the same production as a 100-ft thick formation of 10 md, if all other conditions are equivalent.

1.2.7 Transmissibility

Transmissibility is flow capacity divided by viscosity or kh/μ with units of md ft/cp. An increase in either reservoir permeability or thickness or a decrease in fluid viscosity will improve transmissibility of the fluid in the porous system.

1.2.8 Resistivity and Electrical Conductivity

Electrical conductivity, the electrical analog of permeability, is the ability of a material to conduct an electrical current. With the exception of certain clay minerals, reservoir rocks are nonconductors of electricity. Crude oil and gas are also nonconductors. Water is a conductor if dissolved salts are present so the conduction of an electric current in reservoir rocks is due to the movement of dissolved ions in the brine that occupies the pore space. Conductivity varies directly with the ion concentration of the brine. Thus, the electrical properties of a reservoir rock depend on the fluids occupying the pores and the geometry of the pores.

Resistivity, which is the reciprocal of conductivity, defines the ability of a material to conduct electric current:

$$R = \frac{rA}{L} \tag{1.44}$$

where R is the resistivity in ohm-meters, r is the resistance in ohms, A is the cross-sectional area in m², and L is the length of the conductor in meters. As seen in Figure 1.28, resistivity of water varies inversely with salinity and temperature [41].

During flow through a porous medium, the tortuous flow paths cause the flowing fluid to travel an effective length, L_e, that is longer than the

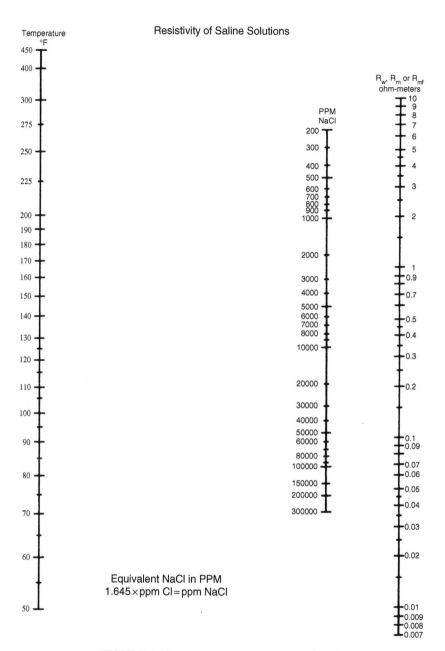

FIGURE 1.28 Resistivity of saline solutions [41].

measured length, L. Some authors have defined this tortuosity, τ, as (L_e/L) while $(L_e/L)^2$ has been used by others.

1.2.9 Formation Resistivity Factor

The formation resistivity factor, F_R, is the ratio of the resistivity of a porous medium that is completely saturated with an ionic brine solution divided by the resistivity of the brine:

$$F_R = \frac{R_o}{R_w} \qquad (1.45)$$

where R_o is the resistivity (ability to impede the flow electric current) of a brine-saturated rock sample in ohm-m, R_w is the resistivity of the saturating brine in ohm-m, and F_R is dimensionless. The formation resistivity factor, which is always greater than one, is a function of the porosity of the rock (amount of brine), pore structure, and pore size distribution. Other variables that affect formation factor include composition of the rock and confining pressure (overburden).

Archie [42] proposed an empirical formula that indicated a power-law dependence of F_R on porosity:

$$F_R = \phi^{-m} \qquad (1.46)$$

where ϕ is porosity and m is a constant (commonly called the cementation factor) related to the pore geometry. The constant, m, was the slope obtained from a plot of F_R vs. porosity on log-log paper. For consolidated, shale free sandstones, the value of m ranged from 1.8 to 2. For clean, unconsolidated sands, m was found to be 1.3, and Archie speculated that m might vary from 1.3 to 2 for loosely or partly consolidated sands. Equations 1.45 and 1.46 were also combined by Archie to give:

$$R_o = R_w \phi^{-m} \qquad (1.47)$$

so that a reasonable estimate of F_R or R_o can be made if the slope, m, is obtained.

Several other correlations [43–55], mostly empirical, between formation factor and porosity have been reported in the literature and these are summarized in Table 1.7.

From an analysis of about 30 sandstone cores from a number of different reservoirs throughout the United States, Winsauer et al. [45] presented what is now known as the Humble relation:

$$F_R = 0.62 \phi^{-2.15} \qquad (1.48)$$

Wyllie and Gregory [46], citing their results and the results of Winsauer et al. [45], proposed that the data for consolidated porous media could be best described empirically by:

$$F_R = a \phi^{-m} \qquad (1.49)$$

TABLE 1.7   Correlations of Formation Resistivity Factor and Porosity

Source	Relation	Notes
Archie [42]	$F_R = \phi^{-m}$	For consolidated sands, m = 1.8 to 2.5. For unconsolidated sands, m = 1.3
Wyllie and Rose [43]	$F_R = \dfrac{\tau^{1/2}}{\phi}$	Tortuosity $\tau = L_e/L$
Tixier [44]	$F_R = \dfrac{1}{\phi^2}$	For limestone
Winsauer et al. [45]	$F_R = \dfrac{\tau^2}{\phi}$	Theory
	$F_R = \dfrac{\tau^{1.67}}{\phi}$	Experimental (transport time of flowing ions)
	$F_R = 0.62\ \phi^{-2.15}$	Sandstones containing varying amounts of clay
Wyllie/Gregory [46]	$F_R = a\phi^{-m}$	General form of Archie relation
Cornell and Katz [47]	$F_R = \dfrac{L_e^2}{\phi L}$	F_R directly proportional to length and inversely proportional to area
Owen [48]	$F_R = 0.68\ \phi^{-2.23}$	Logs in dolomite, mud filtrate same resistivity as connate water
Hill and Milburn [49]	$F_R = 1.4\phi^{-1.78}$	Results of 450 sandstone and limestone cores with R_w of 0.01 ohm-m
	$F_R = \phi^{-1.93}$	When a = 1
Wyllie/Gardner [50]	$F_R = \dfrac{1}{\phi^2}$	Model of capillary bundle, for conducting wetting phase
Sweeney/Jennings [51]	$F_R = \phi^{-m}$	25 various carbonates
	m = 1.57	Water-wet
	m = 1.92	Intermediate wettability
	m = 2.01	Oil-wet
Carothers [52]	$F_R = 1.45\ \phi^{-1.54}$	Sandstones
	$F_R = 0.85\ \phi^{-2.14}$	Limestones
Porter/Carothers [53]	$F_R = 2.45\ \phi^{-1.08}$	From California logs
	$F_R = 1.97\ \phi^{-1.29}$	From Gulf Coast logs. All sandstones, $S_w = 100\%$
Timur [54]	$F_R = 1.13\ \phi^{-1.73}$	Analysis of over 1,800 sandstone samples
Perez-Rosales [55]	$F_R = 1 + G(\phi^{-m} - 1)$	General theoretical relation
	$F_R = 1 + 1.03(\phi^{-1.73} - 1)$	Theoretical relation for sandstones

Results of a logging study in the Brown Dolomite formation, in which the resistivity of the mud filtrate was the same as the connate water, were used by Owen [48] to establish a relationship between the true formation factor and porosity. Shown in Figure 1.29 are the relationships obtained with the equations of Archie, Winsauer, et al., and Tixier.

Hill and Milburn [49] provided data from 450 core samples taken from six different sandstone formations and four different limestone reservoirs. The sandstone formations were described as ranging from clean to very shaly.

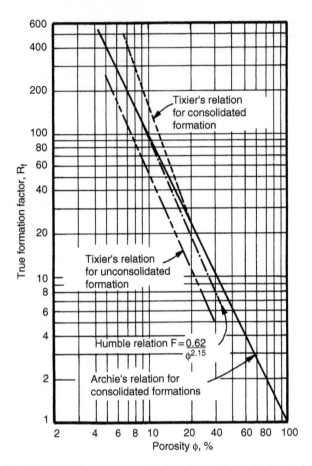

FIGURE 1.29 Formation factor vs. porosity from logs of Brown Dolomite formation [48].

The formation factor was determined at a water resistivity of 0.01 ohm-m since, at that value, the apparent formation factor approached the true formation factor when the rock contained low-resistivity water. From their data, the following equation was provided:

$$F_R = 1.4\phi^{-1.78} \tag{1.50}$$

An expression was also provided for the case in which the constant, a, in Equation 1.49 is taken as unity:

$$F_R = \phi^{-1.93} \tag{1.51}$$

Using 981 core samples (793 sandstone and 188 carbonate), Carothers [52] established a relationship for sands:

$$F_R = 1.45\phi^{-1.54} \tag{1.52}$$

and for limestones:

$$F_R = 0.85\phi^{-2.14} \tag{1.53}$$

As shown in Figure 1.30, a relationship was suggested for calcareous sands:

$$F_R = 1.45\phi^{-1.33} \tag{1.54}$$

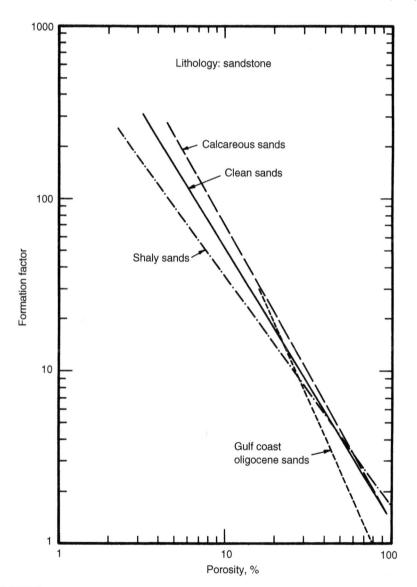

FIGURE 1.30 Formation factor vs. porosity for clean sands, shaly sands, and calcareous sands [52].

and for shaly sands:

$$F_R = 1.65\phi^{-1.33} \tag{1.55}$$

Using these data, the nomograph in Figure 1.31 was constructed to solve the modified Archie expression (Equation 1.49) when it is desired to vary both constants.

Using 1,575 formation factors from California Pliocene well logs, Porter and Carothers [53] presented an in-situ relation:

$$F_R = 2.45\phi^{-1.08} \tag{1.56}$$

and a similar relation for 720 formation factors from Texas-Louisiana Gulf Coast logs:

$$F_R = 1.97\phi^{-1.29} \tag{1.57}$$

This investigation used well log data from sandstone formations known to have water saturations of 100%.

These samples included 569 core plugs (from Alaska, California, Louisiana, Colorado, Trinidad, Australia, and the Middle East) plus 28 samples from Winsauer et al. [45], 362 samples from Hill and Milburn [49], 788 from Carothers [52], and 85 samples from other sources [54].

In another paper [55], Perez-Rosales presented the following theoretical expression:

$$F_R = 1 + G(\phi^{-m} - 1) \tag{1.58}$$

and a generalized equation for sandstones:

$$F_R = 1 + 1.03(\phi^{-1.73} - 1) \tag{1.59}$$

Perez-Rosales notes that the previous expressions are fundamentally incorrect since they do not satisfy the requirement that $F_R = 1$ when $\phi = 1$. A graphical comparison of expressions, provided by Perez-Rosales, is shown in Figure 1.32 for Equations 1.48, 1.59, and 1.60. In porosity ranges of practical interest, the three expressions yield similar results.

From an analysis of over 1,800 sandstone samples, Timur et al. [54] presented the following expression:

$$F_R = 1.13\phi^{-1.73} \tag{1.60}$$

Coates and Dumanoir [56] listed values for the cementation exponent of the Archie equation for 36 different formations in the United States. These data are presented in the following section under "Resistivity Ratio."

In the absence of laboratory data, different opinions have existed regarding the appropriate empirical relationship. Some authors [57] felt that the Archie equation (Equation 1.46) with $m = 2$ or the Humble equation (Equation 1.48) yields results satisfactory for most engineering purposes,

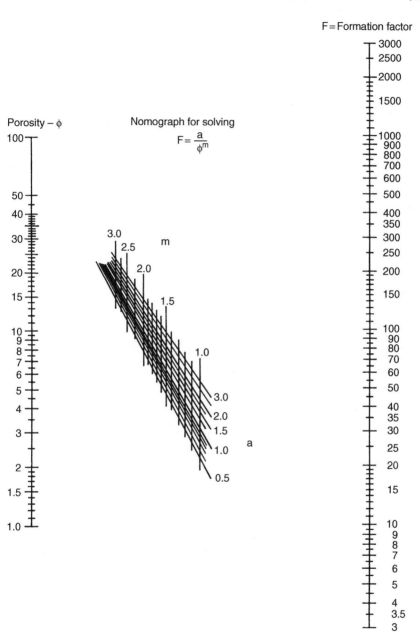

FIGURE 1.31 Nomograph for solving $F_R = a/\phi^m$ [52].

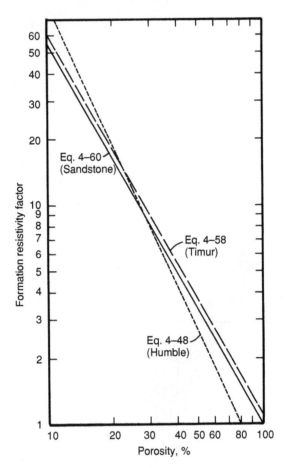

FIGURE 1.32   Graphical comparison of relationship between formation resistivity factor and porosity [55].

but Equation 1.50 may be more valid (these authors point out that the relationship used should be based on independent observations of interest). Another opinion was that, while the Humble relation is satisfactory for sucrosic rocks and the Archie equation with m = 2 is acceptable for chalky rocks, in the case of compact or oolicastic rocks the cementation exponent in the Archie equation may vary from 2.2 to 2.5 [58]. Based on the work of Timur et al. [54], it appears that Equation 1.60 may be more appropriate as a general expression for sandstones, if individual formation factor-porosity relationships are not available for specific cases.

Water in clay materials or ions in clay materials or shale act as a conductor of electrical current and are referred to as conductive solids. Results in Figure 1.33 show that clays contribute to rock conductivity if low-conductivity, fresh, or brackish water is present [59, 60]. The effect of clay

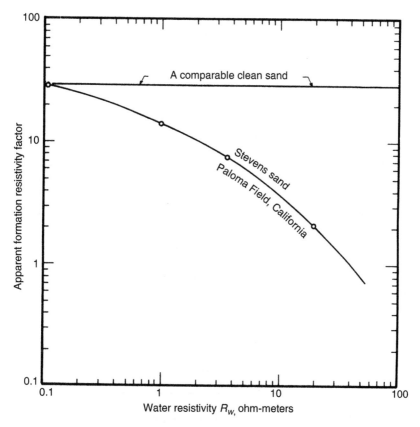

FIGURE 1.33 Effect of clays on formation resistivity factor [59, 60].

TABLE 1.8 Values of Exponent in the Archie Equation for Clays

Mineral particle	Shape factor exponent
Sodium montmorillonite	3.28
Calcium montmorillonite	2.70
Muscovite	2.46
Attapulgite	2.46
Illite	2.11
Kaolinite	1.87

From Reference 61.

on formation resistivity depends on the type, amount, and distribution of clay in the reservoir, as well as the water salinity. Values of m in Equation 1.49 for several clays are given in Table 1.8 [61].

Other variables that affect resistivity of natural reservoir rocks include overburden pressure and temperature during measurement. The value of the cementation exponent, m, is normally higher at overburden conditions [62], especially if porosity is low or with rocks that are not well-cemented. Thus, F_R increases with increasing pressure. Although the effect of temperature depends on clay content of the sample, F_R tends to increase with increasing temperature, but the effect is not as great as pressure [63, 64]. At a fixed pressure, F_R may go through a minimum and then increase as temperature is increased; the combined increase of both temperature and pressure will cause an increase in F_R [64]. Factors that affect the exponent, m, and the coefficient, a, in the modified Archie expression (Equation 1.49) are summarized in Tables 1.9 and 1.10, respectively [65].

To summarize the general relationship between formation resistivity factor and porosity (see Equation 1.49), the normal range for the geometric term, a, is 0.6 to 1.4, and the range for the cementation exponent, m, is 1.7 to 2.5 for most consolidated reservoir rocks [62]. Since the exact values depend on pore geometry and composition of the rock, formation factors should be determined with samples of the reservoir rock of interest, under the reservoir conditions of temperature and overburden pressure.

Based on core analyses of 793 sandstone and 188 carbonate samples, Carothers [52] observed different permeabilities and formation factors for

TABLE 1.9   Factors that Influence the m Exponent in Equation 1.49 for the Rock–Water Interface

1. Pore geometry.
   a. Surface-area-to-volume ratio of the rock particle, angularity, sphericity.
   b. Cementation.
   c. Compaction.
   d. Uniformity of mineral mixture.
2. Anisotropy.
3. Degree of electrical isolation by cementation.
4. The occurrence of an open fracture.

From Reference 65.

TABLE 1.10   Factors that Influence the a Coefficient in Equation 1.49

1. Surface conductance and ionic mobility occurring in water films adsorbed to solid surfaces.
   a. The cation exchange capacity of particular solid materials.
   b. The quantity of water adsorbed to clay particles in the rock framework or within the interstices
2. Salinity of formation water.
3. Wettability relations between particular solid surfaces and hydrocarbons, as they influence cation exchange.
4. The presence and distribution of electrically conductive solid minerals.

From Reference 65.

samples from the same core even though porosity was identical. Further-more, permeability generally decreased as formation factor increased. For permeabilities above 10 md, there appeared to be a relation between for-mation factor, permeability, and lithology. For sandstones, the general rela-tionship was:

$$k = \frac{7 \times 10^8}{F_R^{4.5}} \tag{1.61}$$

and the relation for carbonates was:

$$k = \frac{4 \times 10^8}{F_R^{3.65}} \tag{1.62}$$

Carothers stated that more data are needed to confirm these observa-tions. Any such relation should be used with caution.

1.2.10   Rock Compressibility

The isothermal rock compressibility is defined as the change in volume of the pore volume with respect to a change in pore pressure:

$$c_f = \frac{1}{V_p} \left(\frac{\partial V_p}{\partial p} \right)_T \tag{1.63}$$

where c_f is the formation (rock) compressibility with common units of psi^{-1}, V_p is pore volume, p is pressure in psi, and the subscript T denotes that the partial derivative is taken at constant temperature. The effective rock com-pressibility is considered a positive quantity that is additive to fluid com-pressibility; therefore, pore volume decreases as fluid pressure decreases [26, 66]. Since overburden pressure of a reservoir is essentially constant, the differential pressure between the overburden pressure and the pore pres-sure will increase as the reservoir is depleted. Thus, porosity will decrease slightly, on the order of only one-half percent for a 1,000 psi change in inter-nal fluid pressure [17]. For different reservoirs, porosities tend to decrease as overburden pressure (or depth) increases. Therefore, porosity under reser-voir conditions may differ from values determined in the laboratory [67]. For sandstones with 15% to 30% porosity, reservoir porosity was found to be about 1% lower under reservoir conditions; for low porosity limestones, the difference was about 10% [68].

One of the commonly cited correlations between rock compressibility and porosity was developed by Hall [69] (Figure 1.34) for several sandstone

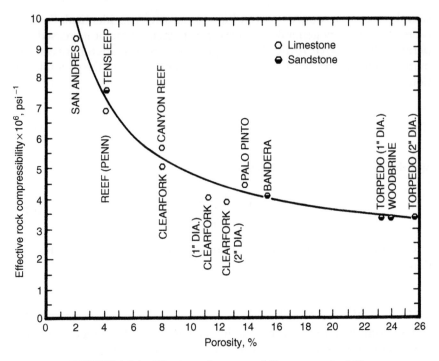

FIGURE 1.34   Effective rock compressibility vs. porosity [69].

and limestone reservoirs. All measurements were conducted with an external pressure of 3,000 psi and internal pressures from 0 to 1,500 psi. Fatt [67] found no correlation between compressibility and porosity, although the porosity range studied (10% to 15%) was very narrow. Van der Knapp [68], citing his measurements and those of Carpenter and Spencer [70], observed a general trend of increasing pore volume compressibility with decreasing porosity. For a particular limestone reservoir, Van der Knapp [68] found that pore compressibility and porosity were related by a simple empirical formula. However, in a more extensive study, Newman [71] suggests that any correlation between pore volume compressibility and porosity does not apply to a wide range of reservoir rocks. As shown in Figure 1.35a, Newman's study in limestones showed poor agreement with the correlations of Hall and Van der Knapp. Figures 1.35b to 1.35d show a comparison of Newman's data with Hall's correlation for consolidated sandstones, friable sandstones, and unconsolidated sandstones. While the general trend of Newman's data on consolidated sandstones (Figure 1.35b) is in the same direction as Hall's correlation, the agreement is again poor. Figure 1.35c shows no correlation for Newman's friable sandstones and Figure 1.35d for unconsolidated sandstones shows an opposite trend from the correlation

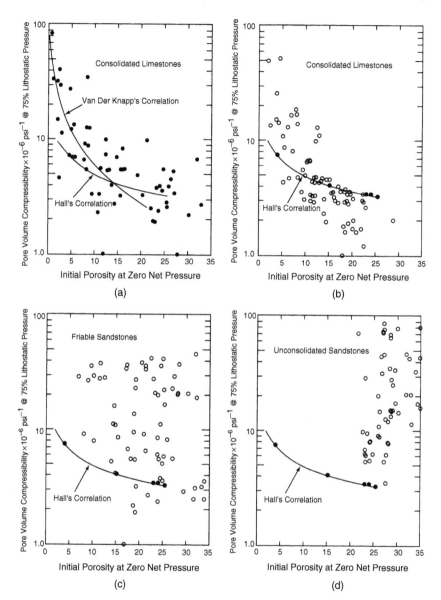

FIGURE 1.35 Pore-volume compressibility at 75% lithostatic pressure vs. sample porosity [71].

presented by Hall. From Newman's data, ranges of compressibilities for various types of reservoir rocks are given in Table 1.11. Clearly, formation compressibility should be measured with samples from the reservoir of interest.

TABLE 1.11   Range of Formation Compressibilities

Formation	Pore volume compressibility, psi^{-1}
Consolidated sandstones	1.5×10^{-6} to 20×10^{-6}
Consolidated limestones	2.0×10^{-6} to 35×10^{-2}
Friable sandstones	2.5×10^{-6} to 45×10^{-6}
Unconsolidated sandstones	5.5×10^{-6} to 85×10^{-6}

From Reference 71.

1.3 PROPERTIES OF ROCKS CONTAINING MULTIPLE FLUIDS

1.3.1 Total Reservoir Compressibility

The total compressibility of oil- or gas-bearing reservoirs represents the combined compressibilities of oil, gas, water, and reservoir rock in terms of volumetric weighting of the phase saturations:

$$c_t = c_o S_o + c_w S_w + c_g S_g + c_f \tag{1.64}$$

where c_t is the total system isothermal compressibility in vol/vol/psi, c_o, c_w, c_g, and c_f are the compressibilities in psi^{-1} of oil, water, gas, and rock (pore volume), respectively, S is fluid saturation, and the subscripts, o, w, and g refer to oil, water, and gas, respectively.

Based on the treatment by Martin [72], Ramey [26] has expressed volumes in terms of formation volume factors with consideration for gas solubility effects:

$$
\begin{aligned}
c_t = S_o &\left[\frac{B_g}{B_o} \left(\frac{\partial R_s}{\partial p} \right) - \frac{1}{B_o} \left(\frac{\partial B_o}{\partial p} \right) \right] \\
+ S_w &\left[\frac{B_g}{B_w} \left(\frac{\partial R_{sw}}{\partial p} \right) - \frac{1}{B_w} \left(\frac{\partial B_w}{\partial p} \right) \right] \\
+ S_g &\left[-\frac{1}{B_g} \left(\frac{\partial B_g}{\partial p} \right) \right] + c_f
\end{aligned}
\tag{1.65}
$$

where p is pressure in psi, R_s is the solubility of gas in oil in scf/STB oil, R_{sw} is the solubility of gas in water in scf/STB water, and B_g, B_o, and B_w are the formation volume factors of gas, oil, and water, respectively.

Fluid and rock compressibilities have been discussed in prior sections of this chapter. Table 1.12 provides a summary of these data.

The rock compressibilities in Table 1.12 represent a majority of the consolidated sandstone and limestone data from Newman [71] that have porosities in the range of 10% to 30%. Oil compressibility increases as a function

TABLE 1.12 Summary of Compressibility Values

	Compressibility, psi^{-1}	
	Range	**Typical value**
Consolidated rock*	2×10^{-6} to 7×10^{-6}	3×10^{-6}
Oil [17, 73]	5×10^{-6} to 100×10^{-6}	10×10^{-6}
Water (gas-free) [26]	2×10^{-6} to 4×10^{-6}	3×10^{-6}

	Compressibility, psi^{-1}	
	At 1,000 psi	**At 5,000 psi**
Gas [26]	$1,000 \times 10^{-6}$	100×10^{-6}
Water (with dissolved gas) [26]	15×10^{-6}	5×10^{-6}

*See Figure 1.35 (for most of samples having porosities of $20 \pm 10\%$ in Figures 1.35a and 1.35b).

of increasing API gravity, quantity of solution gas, or temperature [17]. As pointed out by Ramey [26], when the magnitude of water compressibility is important, the effect of solution gas in the water will be more important. Clearly, the magnitude of gas compressibility will dominate the total system compressibility if gas saturations are high.

In many gas reservoirs, only the gas terms in Equation 1.64 may be significant so that the total system compressibility becomes [26]:

$$c_t = c_g S_g \qquad (1.66)$$

In certain cases of high pressure and high water saturation, rock and water compressibility may be significant so that Equation 1.65 must be used [26].

In oil reservoirs, gas saturations may be low and, even though gas compressibility is much larger than the other compressibilities, each term in Equation 1.64 or 1.65 should normally be considered [26]. In some cases, not all of the compressibility terms will be important. For example, if reservoir pressure is above the saturation pressure, the gas saturation will be zero [20]. However, if the gas saturation exceeds 2% or 3%, the gas compressibility term dominates the total system compressibility and the other terms become insignificant [20].

1.3.2 Resistivity Index

Since crude oil and natural gas are nonconductors of electricity, their presence in reservoir rock increases resistivity. The resistivity index or ratio, I, is commonly used to characterize reservoir rocks that are partially saturated with water and also contain oil or gas, or both:

$$I = \frac{R_t}{R_o} \qquad (1.67)$$

where R_t is the resistivity of the rock at some condition of partial water saturation, S_w, and R_o is the resistivity of the rock when completely saturated with water or brine.

Citing the work of Martin et al. [74], Jakosky and Hopper [75], Wyckoff and Botset [76], and Leverett [77], in which the variation in resistivity with water saturation was studied, Archie [42] plotted the resistivity ratio versus S_w on log-log paper (Figure 1.36). For water saturations down to about 0.15 or 0.20, the following approximate equation appeared to hold, regardless of whether oil or gas was the nonconducting fluid:

$$S_w = \left(\frac{R_o}{R_t}\right)^{1/n} \tag{1.68}$$

where n has been commonly referred to as the saturation exponent. For clean sands and for consolidated sandstones, the value of n was close to 2.0, so the approximate relation was given by Archie as:

$$S_w = \left(\frac{R_o}{R_t}\right)^{1/2} \tag{1.69}$$

By substituting the equation for R_o (refer to Equation 1.47), Archie presented the relationship between water saturation, formation resistivity

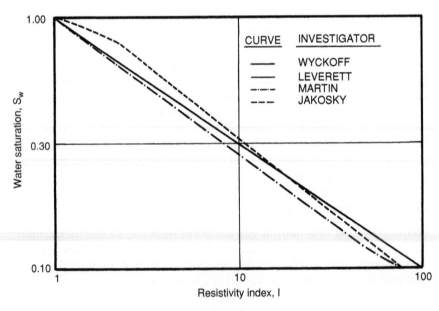

FIGURE 1.36   Variation of resistivity index with water saturation [42].

factor, brine resistivity, and the resistivity of the rock at the given S_w:

$$S_w = \left(\frac{F_R R_w}{R_t}\right)^{1/2}$$ (1.70)

The more general form of Equation 1.70, commonly used, is:

$$S_w = \left(\frac{F_R R_w}{R_t}\right)^{1/n} = \left(\frac{R_w \phi^{-m}}{R_t}\right)^{1/n}$$ (1.71)

An even more general form recognizes that the constant, a, in Equation 1.49 is not necessarily unity:

$$S_w = \left(\frac{a R_w}{\phi^m R_t}\right)^{1/n}$$ (1.72)

The foregoing equations are close approximations in clean formations having a regular distribution of porosity. The accuracy of the equations will not be as good in formations with vugs or fractures.

As shown in Figure 1.37, Patnode and Wyllie [59] found that the presence of clays affected the relationship between water saturation and resistivity

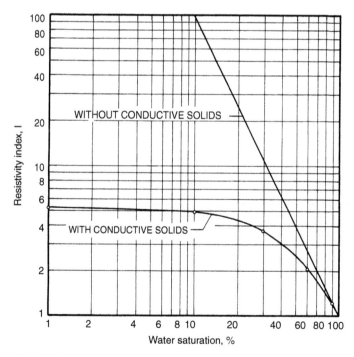

FIGURE 1.37 Effect of conductive solids on the relationship between resistivity index and water saturation [59].

index. As the water saturation is reduced toward zero, the resistivity approaches that of the clays rather than approaching infinity as with clean sands. Relationships between resistivities and fluid content in the presence of conductive solids have been presented in the literature [78, 79].

Early investigations, using data from the Woodbine sand of east Texas, suggested that the saturation exponent, n, may range from 2.3 to 2.7 [80, 81]. Wyllie and Spangler [82] presented data (Figure 1.38) for several natural and synthetic porous media that showed a variation of the saturation exponent from 1.4 to 2.5. Other investigators found that the distribution of fluids within the core sample, at the same water saturation, could affect the resistivity index for both sandstones [83] and for limestones [84]. The exponent, n, was also found to vary depending upon the manner in which the conducting wetting phase saturation was varied [82, 85].

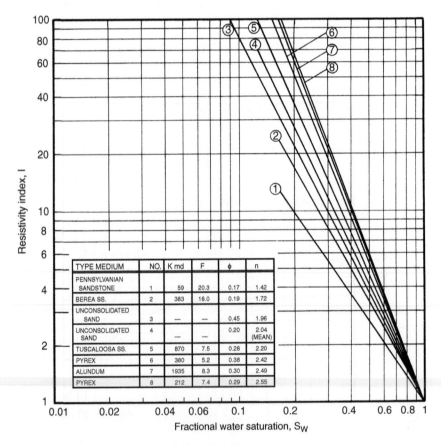

FIGURE 1.38 Relationship between resistivity index and water saturation for several media [82].

Data [56] from a number of different reservoirs for the values of m, the cementation or shape factor, and n, the saturation exponent, in Equation 1.71 are presented in Table 1.13.

Based on these data and other reasoning, Coats and Dumanoir [56] have proposed that the two exponents, m and n, can be assumed to be equal for water-wet, consolidated reservoirs. Ransome [65] has proposed that the saturation exponent may be a special case of the porosity exponent, and the two exponents may bear certain similarities but not necessarily the same value.

TABLE 1.13   Values of Constants in Equation 1.71

	Lithology	Avg. m	Avg. n
Ordovician Simpson, W. Texas and E. New Mexico	SS	1.6	1.6
Permian, W. Texas	SS	1.8	1.9
Ellenburger, W. Texas	LS and Dol.	2.0	3.8
Pennsylvanian, W. Texas	LS	1.9	1.8
Viola, Bowie Field, No. Texas	LS	1.77	1.15
Edwards, So. Texas	LS	2.0	2.8
Edwards Lime, Darst Creek, So. Texas	LS	1.94, 2.02	2.04, 2.08
Frio, So. Texas	SS	1.8	1.8
Frio, Agua Dulce, So. Texas	SS	1.71	1.66
Frio, Edinburgh, So. Texas	SS	1.82	1.47, 1.52
Frio, Hollow Tree, So. Texas	SS	1.80, 1.87	1.64, 1.69
Government Wells, So. Texas	SS	1.7	1.9
Jackson, Cole Sd., So. Texas	SS	2.01	1.66
Miocene, So. Texas	Cons. SS	1.95	2.1
	Uncons. SS	1.6	2.1
Navarro, Olmos, Delmonte, So. Texas	SS	1.89	1.49
Rodessa, E. Texas	LS	2.0	1.6
Woodbine, E. Texas	SS	2.0	2.5
Travis Peak and Cotton Valley, E. Texas and W. Louisiana	HD. SS	1.8	1.7
Wilcox, Gulf Coast	SS	1.9	1.8
Annona, No. Louisiana	Chalk	2.0	1.5
Cockfield, So. Louisiana	SS	1.8	2.1
Frio, Chocolate Bayou, Louisiana	SS	1.55–1.94	1.73–2.22
Sparta, So. Louisiana (Opelousas)	SS	1.9	1.6
Nacatoch, Arkansas	SS	1.9	1.3
Pennsylvanian, Oklahoma	SS	1.8	1.8
Bartlesville, Kansas	SS	2.0	1.9
Simpson, Kansas	SS	1.75	1.3
Muddy, Nebraska	SS	1.7	2.0
Lakota Sd., Crook Co., Wyoming	SS	1.52	1.28
Madison, No. Dakota	LS	1.9	1.7
Mississippian, Illinois	LS	1.9	2.0
Mississippian, Illinois	SS	1.8	1.9

After Reference 56.

As pointed out by Dorfman [86], data in Table 1.13 strongly suggest that assuming a saturation exponent of 2 can result in serious errors in the estimation of water saturation. In low-porosity formations, such as the Cotton Valley sandstone, the saturation was found to vary greatly from the value of 2 [87]. If n is always assumed to be 2, Dorfman contends that many hydrocarbon zones will be overlooked and many water-producing zones could be tested. As related by Hilchie [88], most of the values for the saturation exponent have been obtained at atmospheric conditions and there is the need to obtain laboratory measurements under simulated reservoir pressure and temperature. At atmospheric pressure, the percentage of smaller pores is larger than at reservoir pressure [64], which results in the wrong saturation exponent and a higher value of water saturation [88].

1.3.3 Surface and Interfacial Tensions

The term *interface* indicates a boundary or dividing line between two immiscible phases. Types of interfaces include: liquid-gas, liquid-liquid, liquid-solid, solid-gas, and solid-solid. For fluids, molecular interactions at the interface result in a measurable tension which, if constant, is equal to the surface free energy required to form a unit area of interface. For the case of a liquid which is in contact with air or the vapor of that liquid, the force per unit length required to create a unit surface area is usually referred to as the surface tension. Interfacial tension is used to describe this quantity for two liquids or for a liquid and a solid. Interfacial tension between two immiscible liquids is normally less than the surface tension of the liquid with the higher tension, and often is intermediate between the individual surface tensions of the two liquids of interest. Common units of surface or interfacial tension are dynes per centimeter (or the identical ergs/cm^2) with metric units in the equivalent milli-Newton per meter (mN/m).

The surface tension of pure water ranges from 72.5 dynes/cm at 70°F to 60.1 dynes/cm at 200°F in an almost linear fashion with a gradient of 0.095 dynes/cm/°F [25]. Salts in oilfield brines tend to increase surface tension, but surface active agents that may dissolve into the water from the oil can lower surface tension. At standard conditions, surface tensions of brines range from 59 to 76 dynes/cm [25]. As shown in Figure 1.39, dissolved natural gas reduces surface tension of water as a function of saturation pressure [89].

At a given temperature, surface tension of hydrocarbons in equilibrium with the atmosphere or their own vapor increases with increasing molecular weight (Figure 1.40) [90]. For a given hydrocarbon, surface tension decreases with increasing temperature. At 70°F, surface tensions of crude oils often range from 24 to 38 dyne/cm [25].

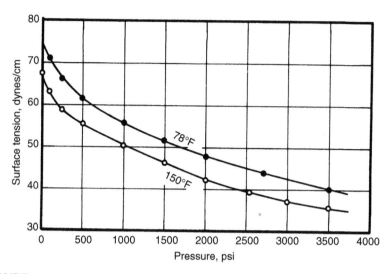

FIGURE 1.39   Surface tension between water and natural gas as a function of saturation pressure [89].

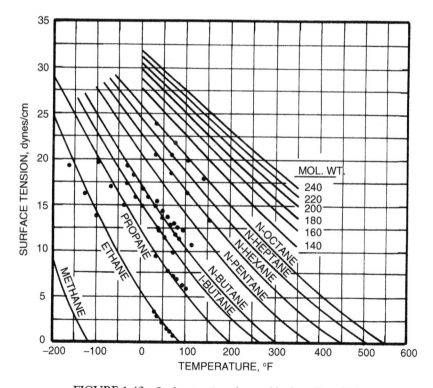

FIGURE 1.40   Surface tension of several hydrocarbons [90].

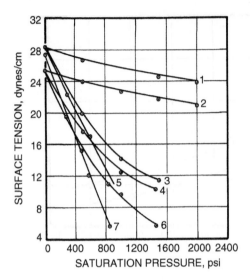

FIGURE 1.41 Surface tension of several crude oils [91].

The presence of dissolved gases greatly reduces surface tension of crude oil as shown in Figure 1.41 [91]. Dissolved natural gas reduces surface tension of crude oil more than previously noted for water, but the amount and nature of gas determines the magnitude of the reduction. The direct effect of a temperature increase on reduction of surface tension more than counterbalances the decreased gas solubility at elevated temperatures. Thus, surface tension at reservoir temperature and pressure may be lower than indicated by Figure 1.41 [25].

Under reservoir conditions, the interfacial interaction between gas and oil involves the surface tension of the oil in equilibrium with the gas. Similarly, the interaction between oil and water determines the interfacial tension between the crude and brine. Listed in Table 1.14 are the surface and interfacial tensions for fluids from several Texas fields [92].

The effect of temperature on interfacial tensions for some oil-water systems is shown in Figure 1.42 [92]; the reduction in interfacial tension with increasing temperature is usually somewhat more pronounced than is observed for surface tension. Although no quantitative relation is observed, the general trend suggests lower interfacial tensions for the higher API gravity crudes. However, in studies with a crude oil containing large amounts of resins and asphaltenes, different effects of temperature on interfacial tension were observed when measurements made at aerobic conditions were compared to anaerobic tests [93]. Interfacial tension between the crude and reservoir brine showed a decrease with an increase in temperature under aerobic conditions, whereas at anaerobic conditions, interfacial tension increased with increasing temperatures. This difference in

TABLE 1.14   Surface and Interfacial Tensions for Several Texas Fields

Field	Formation	Depth, ft	Oil Gravity, °API	Surface Tension, Dynes/cm Oil	Water	Oil-Water Interfacial Tension, Dynes/cm 70°F	100°F	130°F
Breckenridge	Marble Falls	3,200	38.2	28.8	67.6	19.0	10.9	
South Bend	Strawn	2,300	36.1	29.9	61.5	29.1	21.4	
Banyon	Austin	2,135	37.0	29.3	72.5	24.4	17.4	9.6
South Bend	Marble Falls	3,900	25.5	31.8	71.4	24.5	16.9	
Banyon	Austin	2,255	37.9	28.9	72.1	16.9	13.6	12.9
Salt Flats	Edwards	2,700	34.9	30.0	73.0	23.0	16.9	16.7
Driscoll	Catahoula	3,929	26.0	32.4	61.4	20.4	16.0	15.5
Wortham	Woodbine	2,800	38.3	29.2	63.2	13.6	7.3	
Wortham	Corsicana	2,200	22.4	33.2	59.6	25.1	16.7	13.2
Mexia	Woodbine	3,000	36.4	30.0	66.2	21.4	19.0	17.6
Powell	Woodbine	3,000	22.9	30.0	66.2	22.6	15.0	
Wortham	Woodbine	2,800	22.2	33.3	66.0	25.8	15.6	
Mexia	Woodbine	3,000	36.6	30.2	66.6	15.0	9.2	
Breckenridge	Marble Falls	3,200	37.7	28.9	70.1	16.2	8.5	10.0
Breckenridge	Marble Falls	3,200	36.6	29.4	74.1	15.5	11.3	
South Bend	Marble Falls	4,200	38.6	28.9	68.1	14.8	10.8	10.1
Van	Woodbine	2,710	33.9	29.0	61.7	18.1	16.2	
Raccoon Bend	Cockfield	3,007	34.1	31.6	69.8	24.7	14.6	14.3
Tomball	Cockfield	5,541	41.6	28.5	62.0	14.1	13.6	
Van	Woodbine	2,710	35.0	28.8	64.1	17.9	15.0	7.8
Saxet	Catahoula	4,308	26.2	32.0	65.2	17.2	11.5	10.8
Saxet	Catahoula	4,308	27.1	32.3	66.5	20.9	16.5	14.1
Pierce Junction	Frio	4,325	29.4	31.0	62.0	16.9	13.9	8.7
Pierce Junction	Frio	4,335	22.2	32.6	64.1	20.7	12.9	2.1
East Texas	Woodbine	3,660	36.5	28.2	68.6	19.7	10.9	9.6
East Texas	Woodbine	3,660	39.5	27.5	70.2	31.4	17.9	13.9
Goose Creek	Pliocene	1,470	14.2	34.1	63.7	24.4	19.5	
Goose Creek	Pliocene	2,040	21.1	33.6	63.5	18.8	15.3	12.5
Goose Creek	Mio-Pliocene	2,560	21.2	33.3	64.2	18.1	12.9	12.5
Talco	Glen Rose	5,000	23.0	31.9	73.9	20.5	18.8	14.8
Big Lake	Ordovician	8,300	42.6	28.5	63.3	18.1	14.8	12.5
Big Lake	Permian	3,000	38.0	27.9	66.2	27.3	18.3	15.7
Crane	Permian	3,500	31.1	29.5	68.2	18.6	14.8	7.8
Echo	Frye (Pa.)	1,950	38.4	27.8	49.5	34.3	24.6	18.6

From Reference 92.

behavior was attributed to oxidation of the stock tank oil in the aerobic tests. At conditions of reservoir temperature and pressure, interfacial tension of the live reservoir oil was higher than the stock tank oil. The study concluded that live reservoir crude should be used in measurements of interfacial properties and that if stock tank oil is used, at least the temperature should correspond to reservoir conditions.

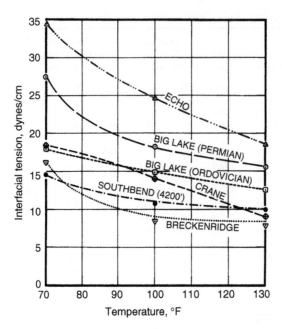

FIGURE 1.42 Influence of temperature on interfacial tensions for crude oil/water systems [92].

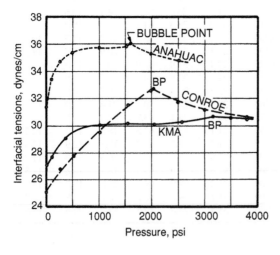

FIGURE 1.43 Effect of dissolved gas and pressure on interfacial tension between crude oil water [89].

Figure 1.43 shows the effect of dissolved gas and pressure on the interfacial tension of three oil-water systems [89]. For each system, interfacial tension increases as the amount of dissolved gas increases, but drops slightly as pressure is increased above the bubblepoint.

Surface and interfacial tensions are important in governing the flow of fluids in the small capillaries present in oil-bearing reservoirs. The capillary forces in oil or gas reservoirs are the result of the combined effect of surface and interfacial tensions, pore size distribution, pore shape, and the wetting properties of the hydrocarbon/rock system.

1.3.4 Wettability and Contact Angle

The contact angle (θ_c), existing between two fluids in contact with a solid and measured through the more dense phase, is a measure of the relative wetting or spreading by a fluid on a solid. A contact angle of zero indicates complete wetting by the more dense phase, an angle of 180° indicates complete wetting of the less dense phase, and an angle of 90° means that neither fluid preferentially wets the solid.

From a combination of Dupre's equation for wetting tension and Young's equation [94], the adhesion tension (τ_A) can be given as [19, 95, 97]:

$$\tau_A = \sigma_{os} - \sigma_{ws} = \sigma_{wo} \cos\theta_{wo} \tag{1.73}$$

where σ_{os} is the interfacial tension between the solid and the less dense fluid phase, σ_{ws} is the interfacial tension between the solid and the more dense phase, and σ_{wo} is the interfacial tension between the fluids of interest. With gas-oil systems, oil is the more dense phase and is always the wetting phase [96]. With oil-water systems water is almost always the more dense phase, but either can be the wetting phase. For oil and water, a positive adhesion tension ($\theta_c > 90°$) indicates a preferentially water-wet surface, whereas a negative adhesion tension ($\theta_c < 90°$) indicates a preferentially oil-wet surface. For a contact angle of 90°, an adhesion tension of zero indicates that neither fluid preferentially wets the solid. Examples of various contact angles are depicted in Figure 1.44 [96].

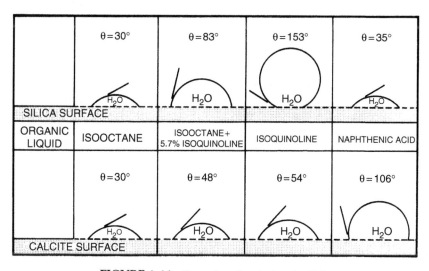

FIGURE 1.44   Examples of contact angles [96].

The importance of wettability on crude oil recovery has been recognized for many years. This subject is discussed in a subsequent section of this chapter. Although Nutting [98] observed that some producing formations were oil-wet, many early workers considered most oil reservoirs to be water-wet (e.g., References [23, 99, 100]; discussion and comments in Reference [96]). From a thermodynamic standpoint, it was felt that pure, clean silica must be wetted by water in preference to any hydrocarbon. In one study [101], no crude oils were tested that had a greater adhesion than pure water. Other results [102] tended to support this contention: capillary pressure tests suggested that all cores tested were water-wet with contact angles ranging from 31° to 80°. However, there are two reasons why these results were obtained [103]: (1) the cores were extracted with chloroform prior to the tests which could have affected the natural wettability, and (2) only receding (decrease in wetting phase saturation) contact angles were measured during the capillary pressure tests. As with capillary pressures, there is a hysteresis in the receding and advancing (increase in wetting phase saturation) contact angles; receding angles are smaller than advancing angles [97]. Bartell and coworkers [95–97] were among some of the first investigators to measure contact angles with crude oil systems that suggested the possibility that oil reservoirs may not be water-wet. Furthermore, they concluded that spontaneous displacement of oil by water should occur only when both advancing and receding angles are less than 90°, and no spontaneous imbibition should occur if the two angles are on opposite sides of 90° [97].

A common technique for measuring advancing contact angles using polished mineral crystals is described in the literature [104]. For many crude oil systems, a considerable amount of time may be required before an equilibrium contact angle on a pure mineral is obtained. As shown in Figure 1.45 [104], some systems that initially appear to be preferentially water-wet become more oil-wet. Small amounts of polar compounds in some crude oils can adsorb on the rock surfaces and change wettability from preferentially water-wet to more oil-wet [96]. A detailed study on how crude oil components affect rock wettability has been made by Denekas et al. [105]. Imbibition tests have been described to examine wettability of reservoir cores [106, 107]. A preferred technique of inferring reservoir wetting from core samples is the Amott method [108] which involves spontaneous imbibition and forced displacement tests; ratios of spontaneous displacement volumes to total displacement volumes are used as an index of wettability. Based on the correlation suggested by Gatenby and Marsden [109], Donaldson et al. [110–112] developed a quantitative indication of wettability, called the U.S. Bureau of Mines (USBM) test, by measuring the areas under capillary pressure curves. The USBM method has the advantage of working well in the intermediate wettability region. A multitude of techniques for the qualitative indication of wettability that

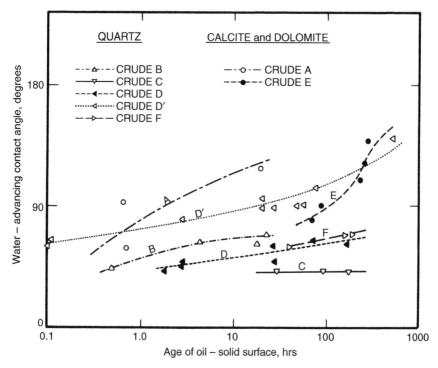

FIGURE 1.45 Change in contact angles with time [104].

have been proposed will not be described but have been discussed in the literature [133].

In a fairly extensive examination of 55 different reservoirs, Treiber, Archer, and Owens [114] arbitrarily assigned water-wet conditions for contact angles of 0° to 75° and oil-wet conditions for contact angles of 105° to 180° with contact angles of 75° to 105° representing an intermediate (referred to as neutral by others) wettability. With these designations, 27% of the samples were water-wet, 66% were oil-wet, and the remaining 7% were of intermediate wettability. Subsequently, Morrow [115] has defined an intermediate wettability when neither fluid spontaneously imbibes in a "squatters' rights" situation. Morrow found that for contact angles less than 62°, the wetting phase would spontaneously imbibe, and for contact angles above 133°, the nonwetting phase would spontaneously imbibe; therefore, the intermediate wettability condition would be operative for contact angles from 62° to 133°. Using Morrow's guidelines, the data of Treiber, Archer, and Owens indicate that 47% of the samples were of intermediate wettability, 27% were oil-wet, and 26% were water-wet. The distribution in wettability according to lithology is given in Table 1.15. In either case, it is apparent that a majority of the samples were not water-wet.

TABLE 1.15   Wettability of 55 Reservoir Rock Samples

	Water-Wet ($\theta_c < 75°$)*	Intermediate ($\theta_c = 75° - 105°$)*	Oil-Wet ($\theta_c > 105°$)*
Sandstones	43%	7%	50%
Carbonate	8%	8%	84%
Total samples	27%	7%	66%
	Water-wet ($\theta_c < 62°$)**	Intermediate ($\theta_c = 62° - 133°$)**	Oil-wet ($\theta_c > 133°$)**
Total samples**	26%	47%	27%

*Based on contact angle/wettability relation suggested by Treiber, Archer, and Owens [114].
**Based on contact angle/wettability relation suggested by Morrow [115].

Using 161 core samples representing various carbonate reservoirs, Chilingar and Yen [116] found that 8% were water-wet ($\theta_c > 80°$), 12% were intermediate ($\theta_c = 80° - 100°$), 65% were oil-wet ($\theta_c = 100° - 160°$), and 15% were strongly oil-wet ($\theta_c = 160° - 180°$). The arbitrary definitions of wettability differ from Trieber et al. [114] and Morrow [115], but the distributions appear to be similar to the carbonate data in Table 1.15.

In the previous discussion, it was implied that pore surfaces within a reservoir rock are uniformly wetted. The concept whereby a portion of the reservoir surfaces are preferentially oil-wet while other portions are preferentially water-wet was termed fractional wettability by Brown and Fatt [117] and Fatt and Klikoff [118]. Fractional wetting was believed to be a result of the varying amount of adsorption of crude oil components on the different minerals present in a reservoir. Other evidence [119, 120] supported the existence of a heterogenerous wetting (also called spotted or Dalmation wettability in the literature). Salathiel [121] introduced the concept of a mixed-wettability condition, a special case of fractional wetting, in which the fine pores and grain contacts are preferentially water-wet and the large pore surfaces are strongly oil-wet. Salathiel concluded that the oil-wet paths can be continuous to provide a means for oil to flow even at very low oil saturations; these results were offered to explain the very good oil recovery noted in some field projects. Then in 1986, Morrow, Lim, and Ward [122] introduced the concept of a speckled wettability in which a rationale is presented whereby oil tends to be trapped in pore throats rather than pore bodies. Speckled wettability mimics behavior of strongly water-wet conditions observed during waterflooding: water breakthrough is abrupt, relative permeability to water at residual oil saturation is low, water is imbibed spontaneously, and oil is not imbibed spontaneously.

When cores are obtained for laboratory tests where wettability is important, precautions must be taken to ensure the wetting preference of the formation is not altered during coring. Mud additives, such as dispersants, weighting agents, lost circulation materials, thinners or colloids, that

TABLE 1.16   Effect of Water-Base Mud Additives on the Wettability of Cores

Wettability of Test Cores after Exposure to Filtrate

Component	Water-Wet Limestone	Water-Wet Sandstone	Oil-Wet Sandstone
Rock-salt	No change	No change	No change
Starch	Slightly less water-wet	Slightly less water-wet	—
CMC	No change	No change	Water-wet
Bentonite	No change	No change	Water-wet
Tetrasodium pyrophosphate	No change	Less water-wet	—
Calcium lignosulfonate	No change	Less water-wet	—
Lime	No change	Slightly more water-wet	Water-wet
Barite	No change	No change	—

From Reference 107.

possess surface-active properties may drastically change core wettability. Surface active agents should be avoided so that the core samples have the same wettability as the reservoir rock. Listed in Table 1.16 are the effects of various mud additives on wettability of water-wet and oil-wet cores [107].

In the case of water-wet sandstone or limestone cores, rock-salt, bentonite, carboxymethylcellulose (CMC), and barite had no effect on wettability. However, oil-wet sandstone cores were reversed to a water-wet condition when exposed to CMC, bentonite, or lime solutions. Additional tests with bentonite solutions indicated that wettability of oil-wet cores is not reversed if the solution pH is lowered to a neutral or slightly acidic value. These results suggest that from a wettability standpoint, the best coring fluid is water (preferably formation brine); if bentonite is used, mud pH should be neutral or slightly acidic. If appreciable hydrogen sulfide is suspected in the interval being cored, it may be undesirable to lower pH. In fact, a very alkaline mud (pH 10–12) may be used to keep the sulfide in the ionized state for safety and corrosion considerations. Subsequent work [123] suggests the preferred system to obtain fresh cores is a natural water-base mud with no additives, or a bland mud consisting of bentonite, salt, and CMC. However, other results [124] conclude that bland muds may not, in fact, be bland. While none of the bland additives altered wettability of water-wet rock samples, all of the components with the exception of bentonite made oil-wet samples significantly less oil-wet. The bland additives that were tested included bentonite, pregelatinized starch, dextrid (an organic polymer), drispac (a polyanionic cellulose polymer), hydroxyethylcellulose, xanthan gum polysaccharide, and CMC. All of the drilling mud components considered to be bland decreased the amount of oil imbibed into a core and increased the amount of water imbibed.

TABLE 1.17   Empirical Relative Permeability Equations

I. *Oil-gas relative permeabilities*
(for drainage cycle relative to oil)

$$S^* = \frac{S_o}{(1-S_{iw})}$$

Where S_{iw} is the irreducible water
saturation.

	k_{ro}	k_{rg}
A. Unconsolidated sand—well sorted	$(S^*)^{3.0}$	$(1-S^*)^3$
B. Unconsolidated sand—poorly sorted	$(S^*)^{3.5}$	$(1-S^*)^2(1-S^{*1.5})$
C. Cemented sand, oolitic lime, and vugular lime	$(S^*)^{4.0}$	$(1-S^*)^{2.0}(1-S^{*2.0})$

II. *Water-oil relative permeabilities*
(for drainage cycle relative to water)

$$S^* = \frac{S_w - S_{iw}}{(1-S_{iw})}$$

where S_{iw} is the irreducible water
saturation.

	k_{ro}	k_{rw}
A. Unconsolidated sand—well sorted	$(1-S^*)^{3.0}$	$(S^*)^{3.0}$
B. Unconsolidated sand—poorly sorted	$(1-S^*)^2(1-S^{*1.5})$	$(S^*)^{3.5}$
C. Cemented sand, oolitic lime, and vugular lime	$(1-S^*)^2(1-S^{*2.0})$	$(S^*)^{4.0}$

From Reference 20.

Preventing wettability changes in core material, after it has been recovered at the surface, is equally important so that subsequent laboratory measurements are representative of reservoir conditions. Changes in wettability of core material that occur during handling or storage are usually caused by oxidation of the crude oil, evaporation of volatile components, or decreases in temperature or pressure which cause the deposition of polar compounds, asphaltenes, or heavy hydrocarbon compounds [107, 125]. Because of the complexity of the mechanisms involved, the magnitude and direction of changes in wetting conditions, when reservoir cores are preserved, are not fully understood. Weathering of water-wet cores has been reported to frequently cause the grain surfaces to become oil-wet [107]. In other experiments [126] oil-wet cores changed to water-wet upon contact with air. Therefore it is necessary to preserve core samples at the well-site to ensure that wettability is not altered by contamination, oxidation, or evaporation. Two methods of preserving conventional cores, immediately after they have been removed from the core barrel, will prevent changes in wettability for several months [107]. One method consists of immersing the core in deoxygenated formation brine or suitable synthetic brine (i.e., drilling mud filtrate) and keeping the sample in suitable nonmetallic containers that can be sealed to prevent leakage and the entrance of oxygen. In a second method, the cores are wrapped in Saran or polyethylene film and aluminium foil, and then coated with wax or strippable plastic. Cores obtained by either of these methods are referred to as preserved, native-state, or fresh cores, and are preferred for many laboratory tests.

For certain laboratory tests, it may be possible to clean reservoir cores with solvents and resaturate with reservoir fluids to restore the original wetting conditions. Details of preparing such restored-state or extracted cores are discussed subsequently in the section "Coring and Core Analysis." The concept of the method is to clean the core thoroughly until it is water-wet, saturate with reservoir brine, flush with crude oil, and age for over 1,000 hours at reservoir temperature.

Regardless of the method of core handling employed, the rock samples used in the laboratory should have a surface state as close as possible to that present in the reservoir. If preserved cores are used, it is essential they be stored under air-free conditions because exposure to air for as little as 6–8 hours can cause water evaporation and other changes in core properties. If extracted cores are used, drying of the cores can be very critical when hydratable minerals, capable of breaking down at low temperatures are present. Contamination from core holders that contain certain types of rubber sleeves can be prevented by using an inner liner of tubular polyethylene film. Because of the instability of many oilfield waters, it is usually desirable to prepare synthetic brines to prevent core plugging caused by deposition of insolubles.

When possible, tests should be made under reservoir conditions of temperature and pressure using live reservoir oil. This is an improvement over room condition techniques where tests are made at atmospheric conditions with refined laboratory oils. Use of the live crude exposes the rock to compounds present in the oil that might influence wettability, and establishes an environment as close as possible to reservoir conditions. Cores evaluated at atmospheric conditions may be more oil-wet than similar tests at reservoir conditions because of the decreased solubility of wettability-altering compounds at lower temperatures and pressures [107, 123]. In a contact angle study [93] with calcium carbonate crystals and a crude oil containing 27.3% resins and 2.2% asphaltenes, a complete reversal from a predominantly oil-wet system at lower temperatures to a predominantly water-wet system at higher temperatures was found. While pressure alone had little effect on the wettability of the system, the study speculated that the addition of gas-in-solution with increasing pressure should favor a more water-wet situation than would be indicated from laboratory tests at atmospheric conditions. Even when all precautions have been taken, there is no absolute assurance that reservoir wettability has been duplicated.

1.3.5 Capillary Pressure

Curvature at an interface between wetting and nonwetting phases causes a pressure difference that is called capillary pressure. This pressure can be viewed as a force per unit area that results from the interaction of surface forces and the geometry of the system.

Based on early work in the nineteenth century of Laplace, Young, and Plateau (e.g., Reference 94), a general expression for capillary pressure, P_c, as a function of interfacial tension, σ, and curvature of the interface is [19]:

$$P_c = \sigma\left(\frac{1}{r_1} + \frac{1}{r_2}\right) \tag{1.74}$$

where r_1 and r_2 are the principal radii of curvature at the interface. These radii are not usually measured, and a mean radius of curvature is given by the capillary pressure and interfacial tension.

For a cylindrical vertical capillary, such as a small tube, the capillary pressure for a spherical interface is [19]:

$$P_c = \frac{2\sigma\cos\theta_c}{r} = gh(\rho_1 - \rho_2) \tag{1.75}$$

where r is the radius of the tube, θ_1 is the contact angle measured through the more dense phase that exists between the fluid and the wall of the tube, g is the gravitational constant, ρ is density, h is column height, and the subscripts refer to the fluids of interest. For a fluid that wets the wall of a capillary tube, the attraction between the fluid and the wall causes the fluid to rise in the tube. The extent of rise in the capillary is proportional to the interfacial tension between the fluids and the cosine of the contact angle and is inversely proportional to the tube radius.

An analogous situation can occur during two-phase flow in a porous medium. For example if capillary forces dominate in a water-wet rock, the existing pressure differential causes flow of the wetting fluid to occur through the smaller capillaries. However, if viscous forces dominate, flow will occur through the larger capillaries (from Pouiselle's law, as a function of the 4th power of the radius).

Figure 1.46 depicts a typical capillary pressure curve for a core sample in which water is the wetting phase. Variation of capillary pressure is plotted as a function of water saturation. Initially, the core is saturated with the wetting phase (water). The nonwetting phase, oil in this case, is used to displace the water. As shown in the figure, a threshold pressure must be overcome before any oil enters the core. The initial (or primary) drainage curve represents the displacement of the wetting phase from 100% saturation to a condition where further increase in capillary pressure causes little or no change in water saturation. This condition is commonly termed the irreducible saturation, S_{iw}. The imbibition curve reflects the displacement

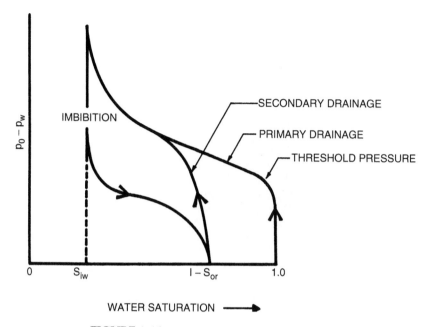

FIGURE 1.46 Example capillary pressure curves.

of the nonwetting phase (oil) from the irreducible water saturation to the residual oil saturation. Secondary drainage is the displacement of the wetting phase from the residual oil saturation to the irreducible water saturation. A hysteresis is always noted between the drainage and imbibition curves. Curves can be obtained within the hysteresis loop by reversing the direction of pressure change at some intermediate point along either the imbibition or secondary drainage curve. The nonuniform cross-section of the pores is the basic cause of the hysteresis in capillary pressure observed in porous media. Therefore, capillary pressure depends on pore geometry, interfacial tension between the fluids, wettability of the system (which will be discussed later in this chapter), and the saturation history in the medium.

Leverett [100] introduced a reduced capillary pressure function (subsequently termed the Leverett J function by Rose and Bruce [127]) that was suggested for correlating capillary pressure data:

$$J(S_w) = \frac{P_c}{\sigma \cos \theta_c} \left(\frac{k}{\phi} \right)^{1/2} \tag{1.76}$$

where $J(S_w)$ = the correlating group consisting of the terms of Equation 1.75

k = the permeability

ϕ = porosity of the sample

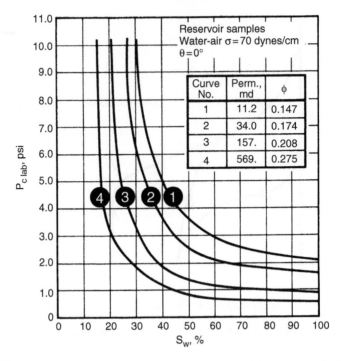

FIGURE 1.47   Capillary pressures of different permeability core samples [20].

The J function was originally proposed as means of converting all capillary pressure data for clean sand to a universal curve. A series of capillary pressure curves are shown as a function of permeability in Figure 1.47 [20]. An example of the J function curve generated from these data is shown in Figure 1.48 [20]. While the J function sometimes correlates capillary pressure data from a specific lithology within the same formation, significant variations can be noted for different formations.

Common laboratory methods of measuring capillary pressure include [19]: mercury injection, porous diaphragm or plate (restored state), centrifuge method, and steady-state flow in a dynamic method. While the restored state test is generally considered the most accurate, mercury injection is routinely used. However, it is necessary to correct the mercury injection data for wetting conditions before comparison to results from the restored state test.

A very valuable use of capillary pressure data is to indicate pore size distribution. Since the interfacial tension and contact angle remain constant during a test such as already described, pore sizes can be obtained from capillary pressures. For rocks with more uniform pore sizes, capillary pressure curves will be close to horizontal. The slope of the capillary pressure curve will generally increase with broader pore-size distribution.

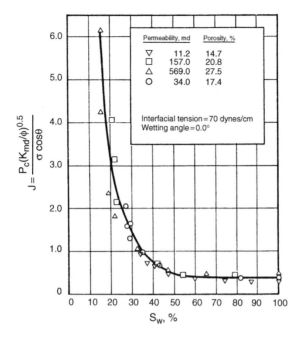

FIGURE 1.48 J function plot for data in Figure 1.47 [20].

If laboratory capillary pressure data are corrected to reservoir conditions, the results can be used for determining fluid saturations. Figure 1.49 shows a close agreement in water saturations obtained from capillary pressure and electric logs [48].

Capillary pressure data are helpful in providing a qualitative assessment of the transition zones in the reservoir. A transition zone is defined as the vertical thickness where saturation changes from 100% water to irreducible water for water-oil contact, or from 100% liquid to an irreducible water saturation for gas-oil contact.

1.3.6 Effective Permeability

In the previous section, "Absolute Permeability," it was stated that permeability at 100% saturation of a fluid (other than gases at low pressure) is a characteristic of the rock and not a function of the flowing fluid. Of course, this implies that there is no interaction between the fluid and the rock (such as interaction between water and mobile or swelling clays). When permeabilities to gases are measured, corrections must be made for gas slippage which occurs when the capillary openings approach the mean free path of the gas. Klinkenberg [128] observed that gas permeability depends on

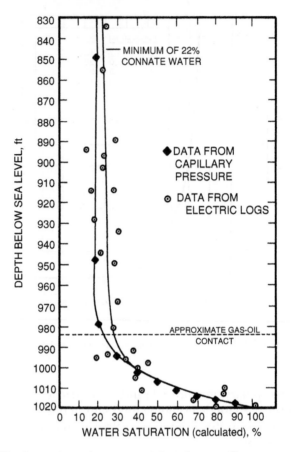

FIGURE 1.49 Comparison of water saturations from capillary pressure and electric logs [48].

the gas composition and is approximately a linear function of the reciprocal mean pressure. Figure 1.50 shows the variation in permeability as a function of mean pressure for hydrogen, nitrogen, and carbon dioxide. Klinkenberg found that by extrapolating all data to infinite mean pressure, the points converged at an equivalent liquid permeability (k_ℓ), which was the same as the permeability of the porous medium to a nonreactive single-phase liquid. From plots of this type, Klinkenberg showed that the equivalent liquid permeability could be obtained from the slope of the data, m, the measured gas permeability, k_g, at a mean flowing pressure $\bar{p}$, at which k_g was observed:

$$k_\ell = \frac{k_g}{1 + (b/\bar{p})} = k_g - \frac{m}{\bar{p}} \qquad (1.77)$$

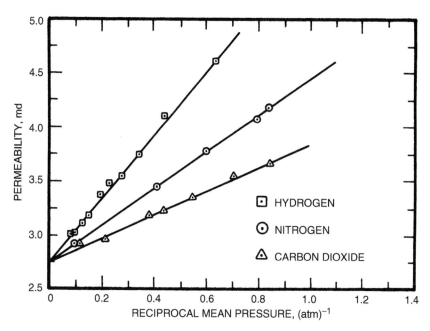

FIGURE 1.50 Gas slippage in core [128].

where b is a constant for a given gas in a given medium and is equal to m divided by k_ℓ. The amount of correction, known as the Klinkenberg effect, varies with permeability and is more significant in low permeability formations.

In studies [129, 130] with very low permeability sandstones, liquid permeabilities were found to be less than gas permeabilities at infinite mean pressure, which is in contrast with the prior results of Klinkenberg. Furthermore, it has been shown [130] that liquid permeabilities decreased with increasing polarity of the liquid. For gas flow or brine flow in low-permeability sandstones, permeabilities were independent of temperature at all levels of confining pressure [130]. The data [130] showed that for a given permeability core sample at a given confining pressure, the Klinkenberg slip factors and slopes of the Klinkenberg plots were proportional to the product of viscosity and the square root of absolute temperature.

As shown in Figure 1.51 permeability of reservoir rocks can decrease when subjected to overburden pressure [131]. When cores are retrieved from a reservoir, the confining forces are removed and the rock can expand in all directions which can increase the dimensions of the available flow paths. In reservoirs where this is significant, it is imperative that permeability measured in the laboratory be conducted at the confining pressure that represents the overburden pressure of the formation tested.

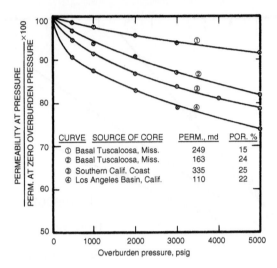

FIGURE 1.51   Air permeability at different overburden pressures [131].

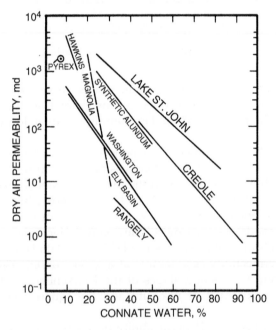

FIGURE 1.52   Relationship between air permeability and connate water saturation [132].

As a general trend, air permeability decreases with increasing connate water saturation. Relationships between air permeability and connate water saturation in Figure 1.52 show a linear decrease in the logarithm of permeability as a function of water saturation that depends on the individual field [132].

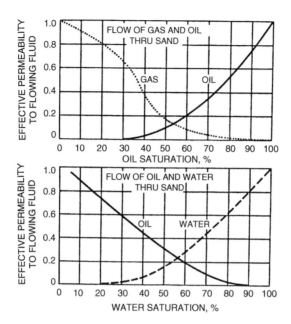

FIGURE 1.53   Effective permeabilities.

The fluid system of an oil reservoir consists usually of gas, oil, and water. In the case of such a heterogeneous system, flow of the different phases is a function of fluid saturation in the reservoir by the different phases. The lower the saturation of a certain liquid, as compared to other liquids, the lower the permeability to that liquid. This type of permeability is termed effective permeability and is defined as permeability of the rock to one liquid under conditions of saturation when more than one liquid is present. Typical permeability-saturation relations for oil and gas and for oil and water are shown in Figure 1.53.

From the practical point of view, permeability may be considered as a measure of productivity of the producing horizon. Knowledge of permeability is useful in a number of reservoir problems. The concept of effective permeability is of particular importance since it emphasizes a need for production practices, which tend to maintain good permeability of the reservoir to oil.

Thus, the absolute permeability is the permeability measured when the medium is completely saturated with a single fluid. Effective permeability is the permeability to a particular fluid when another fluid is also present in the medium. For example, if both oil and water are flowing, the effective permeability to oil is k_o and that to water is k_w. The sum of the effective permeabilities is always less than the absolute permeability [17]. As noted in the previous section, permeability is commonly expressed in millidarcies (md).

1.3.7 Relative Permeability

If the effective permeabilities are divided by a base permeability (i.e., the absolute permeability), the dimensionless ratio is referred to as the relative permeability, namely k_{rg} for gas, k_{ro} for oil, and k_{rw} for water:

$$k_{rg} = \frac{k_g}{k}; \quad k_{ro} = \frac{k_o}{k}; \quad k_{rw} = \frac{k_w}{k} \qquad (1.78)$$

where k_g, k_o, and k_w are the effective permeabilities to gas, oil, and water, respectively, and k is some base permeability that represents the absolute permeability. For gas-oil two-phase relative permeabilities, the base permeability is often the equivalent liquid permeability. For oil-water two-phase relative permeabilities, three different base permeabilities are often used [133]:

1. The permeability to air with only air present.
2. The permeability to water at 100% S_w.
3. The permeability to oil at irreducible water saturation.

Wyckoff and Botset [76] are generally credited with performing the first gas and liquid relative permeabilities which were conducted in unconsolidated sandpacks in 1936. In these early experiments, a relationship was observed between the liquid saturation of a sand and the permeability to a liquid or gas phase [76, 134]. At about the same time, Hassler, Rice, and Leeman [135] measured relative air permeabilities in oil-saturated cores. In 1940, relative permeability measurements were extended to consolidated cores by Botset [136]. Since then, a number of dynamic (fluid displacement or fluid drive) methods [83, 137–143] and static (or stationary-phase) methods [144–150] have been proposed to determine relative permeabilities in core samples. In the latter methods, only the nonwetting phase is allowed to flow by the use of a very low pressure drop across the core; hence, this method is applicable only to the relative permeability of the nonwetting phase. The dynamic methods include: (1) steady-state methods in which fluids are flowed simultaneously through a core sample at a fixed gas-oil or water-oil ratio until equilibrium pressure gradients and saturations are achieved, and (2) unsteady-state methods in which an oil-saturated core is flooded with either gas or water at a fixed pressure drop or flow rate so that the average fluid saturation changes result in a saturation gradient. The most popular steady-state procedure is the Penn State method [83], but the most common dynamic test is the unsteady-state method because of the reduced time requirement. The various methods have been evaluated [139, 151] and generally provide similar results.

Based on the initial work of Leverett [100] and Buckley and Leverett [152], Welge [153] was the first to show how to calculate relative permeability ratios in the absence of gravity effects. Subsequently, Johnson, Bossler, and Naumann [154] showed that each of the relative permeabilities could be calculated even when gravity is not neglected. Other calculations of relative permeabilities have been proposed by Higgins [155], Guerrero and Stewart [156, 157], and a graphical technique has been presented by Jones and Rozelle [158].

An example of water-oil relative permeability plot vs. water saturation is given in Figure 1.54. Several features will be described that pertain to generating relative permeability curves from cores in the laboratory. If a clean, dry core is completely saturated with water, the permeability at 100% S_w should be similar to the equivalent liquid permeability obtained from gas flow measurements at 100% S_g. Exceptions to this generality include some low-permeability systems and other cores that contain clays or minerals that interact with the water used. If a clean core is used, it will probably be strongly water-wet when saturated with brine. As crude oil is injected into the core, the relative permeability to water decreases during the drainage cycle (decreases in wetting phase) while the relative permeability to oil increases. Some water that resides in the nooks and crannies of the pore

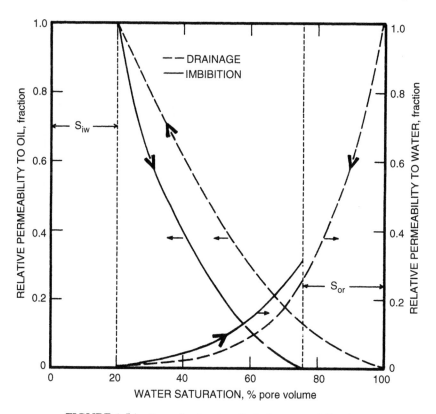

FIGURE 1.54   Example of water-oil relative permeability data.

space cannot be displaced by the oil, regardless of the throughput volume. This water saturation, which does not contribute significantly to occupying the flow paths, is called the irreducible water saturation, S_{iw}. With the core at S_{iw}, there is 100% relative permeability to oil (only oil is flowing) and no permeability to water. At this point, the core can be closed in for about 1,000 hours to allow sufficient time for wettability changes to occur. Then the core is flooded with water in an unsteady-state test, or fixed ratios of water and oil are injected in the steady-state test. If water continues to be the wetting phase, the relative permeability to water (which will be only a function of saturation) will be the same during the drainage and imbibition cycles. (The importance of wettability on relative permeabilities will be discussed in the next section of this chapter.) As the water is injected into the oil-flooded core, k_{rw} increases while k_{ro} decreases. Not all of the oil can be displaced from the core, regardless of the water throughput (at modest flow rates or pressure drops), and this is referred to as the waterflood residual oil saturation, S_{or}.

Similar observations apply to gas-oil relative permeability data as displayed in Figure 1.55. Typically, the gas-oil relative permeabilities are

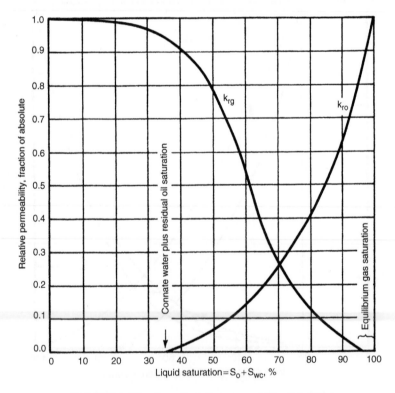

FIGURE 1.55   Gas-oil relative permeability data [20].

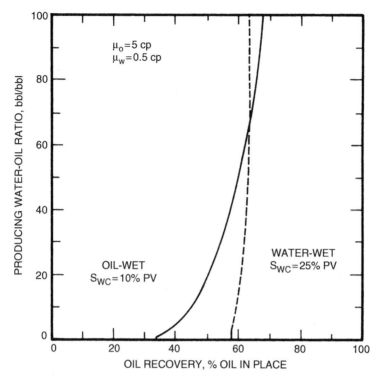

FIGURE 1.56 Comparison of oil recovery for oil-wet and water-wet cores [133].

plotted against the total liquid saturation, which includes not only the oil but also any connate water that may be present. In the presence of gas, the oil (even if connate water is present) will be the wetting phase in preference to gas. As a result, the k_{ro} curve from gas-oil flow tests resembles the drainage k_{rw} curve from oil-water flow tests. As seen in Figure 1.56, the irreducible gas saturation (also called the equilibrium or critical gas saturation) is usually very small. When gas saturation is less than the critical value, gas is not mobile but does impede oil flow and reduces k_{ro}.

Three-phase relative permeabilities pertaining to simultaneous flow of gas, oil, and water have been provided in the literature [19, 50]. Since the occurrence of such three-phase flow in a reservoir is limited [20], relative permeabilities for these conditions will not be discussed in this chapter, and the reader is referred to other sources [19, 20, 137, 140, 159, 160].

Based on the work of Corey [150] and Wyllie [23, 50], empirical equations have been summarized by Slider [20] to estimate relative permeabilities. These equations permit the estimation of k_{ro} from measurements of k_{rg}. Other empirical equations for estimating two-phase relative permeabilities in consolidated rocks are available in the literature [161].

Early work in unconsolidated sands showed that fluid viscosity or the range of permeability had negligible effects on the relationship between relative permeability and fluid saturation [76, 100]. Geffen et al. [141] confirmed that relative permeabilities in cores are not affected by fluid properties provided wettability is not altered. However, others [162, 163] have found that viscosity ratio influences relative permeability data when the displacing fluid is non-wetting. For constant wettability conditions, the higher the viscosity of one of the liquids, the lower is the relative permeability of the other liquid [163].

Geffen et al. [141] did cite a number of factors, in addition to fluid saturations, that affect relative permeability results. Because of capillary hysteresis, saturation history was important in that fluid distribution in the pores was altered. Flow rates during laboratory tests need to be sufficiently high to overcome capillary end effects (retention of the wetting phase at the outlet end of the core) [141]. According to Wyllie [23], relative permeability varies because of varying geometry of the fluid phases present, which is controlled by effective pore size distribution, method of obtaining the saturation (saturation history); heterogeneity of the core sample, and wettability of the rock-fluid system.

Controversy continues to exist regarding the effect of temperature on relative permeabilities (for example see the discussion and prior citations in References 164 and 165). Miller and Ramey [164] observed no change in relative permeability with temperatures for clean systems, and speculated that for reservoir fluid/rock systems, effects such as clay interactions or pore structure would need to be considered. Honarpour et al. [165] summarized the effects of temperature on two-phase relative permeabilities as measured by various researchers. In field situations, the larger overburden pressure associated with greater depths may be more important in affecting relative permeabilities than the associated temperature increases. As with many other tests, relative permeability measurements should be conducted at reservoir conditions of overburden pressure and temperature with crude oil and brine representative of the formation under study.

1.3.8 Effect of Wettability on Fluid-Rock Properties

Oil Recovery and Fluid Saturations   Since a reservoir rock is usually composed of different minerals with many shapes and sizes, the influence of wettability in such systems is difficult to assess fully. Oil recovery at water breakthrough in water-wet cores is much higher than in oil-wet cores [106, 110, 166–169], although the ultimate recovery after extensive flooding by water may be similar, as shown in Figure 1.56. These authors have shown that oil recovery, as a function of water injected, is higher from water-wet cores than from oil-wet cores at economical water-oil ratios. In 1928, Uren and Fahmy [170] observed better recovery of oil from unconsolidated

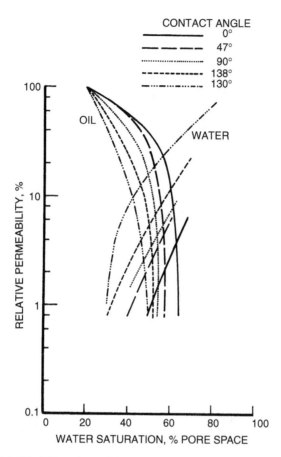

FIGURE 1.57 Effect of wettability on water-oil relative permeabilities [174].

sands that had an intermediate wettability, and several investigators have suggested better oil recovery from cores of intermediate or mixed wettability [106, 108, 121, 123, 171]. Later evidence by Morrow [172] and Melrose and Brandner [173] suggests that mobilization of trapped oil is more difficult in the intermediate wettability region, but the prevention of oil entrapment should be easier for advancing contact angles slightly less than 90°.

With water-wet cores in laboratory waterfloods, the oil production at water breakthrough ceases abruptly and water production increases sharply. With systems that are not water-wet, water breakthrough may occur earlier, but small fractions of oil are produced for long periods of time at high water cuts. In strongly water-wet systems, the residual oil that is permanently trapped by water resides in the larger pores, whereas in oil-wet systems trapping occurs in the smaller capillary spaces [106, 133].

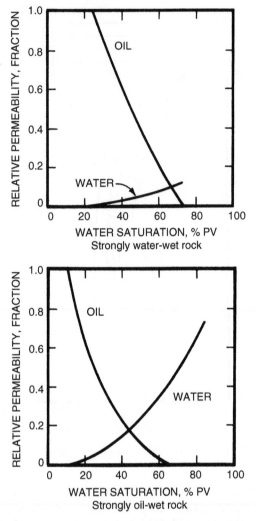

FIGURE 1.58   Strongly water-wet and strongly oil-wet relative permeability curves [133].

Relative Permeability Characteristics For a system having a strong wetting preference for either oil or water, relative permeability of the wetting phase is a function of fluid saturation only [76, 137, 160]. Details of the effect of wettability on relative permeabilities have been discussed by several authors [113, 133, 174, 175]. In a detailed study using fired Torpedo (outcrop sandstone) cores, Owens and Archer [174] changed wettability by adding surface active agents to either the oil or water. Firing of the cores stabilized any clay minerals present and provided more constant internal conditions. Both gas-oil and water-oil relative permeabilities were

TABLE 1.18   Typical Water-Oil Relative Permeability Characteristics

	Strongly Water-Wet	Strongly Oil-Wet
Connate water saturation.	Usually greater than 20% to 25% PV.	Generally less than 15% PV, frequently less than 10%.
Saturation at which oil and water relative permeabilities are equal.	Greater than 50% water saturation.	Less than 50% water saturation.
Relative permeability to water at maximum water saturation; i.e., floodout.	Generally less than 30%.	Greater than 50% and approaching 100%.

From Reference 133.

obtained. Some of the water-oil relative permeability data, all started at the same water saturation and obtained with the Penn State steady-state method, are reproduced in Figure 1.57. As the contact angle was increased to create more oil-wet conditions the effective permeability to oil decreased. Because of the differences in flow paths for the different wettability conditions, oil-wet systems had lower k_{ro} and higher k_{rw} when compared to water-wet conditions at the same water saturation. As the level of oil-wetting increased, k_{ro} at any saturation decreased whereas k_{rw} increased.

Craig [133] has presented typical relative permeability curves, such as given in Figure 1.58, to point out differences in strongly water-wet and strongly oil-wet conditions. Craig suggests that several rules of thumb can help in distinguishing wetting preferences; typical characteristics of water-oil relative permeability curves are given in Table 1.18 [133]. Additionally, the strongly oil-wet relative permeability curves tend to have more curvature. Craig suggests the generality that relative permeability curves of intermediate wettability systems will have some of the characteristics of both water-wet and oil-wet conditions. However, as mentioned earlier in the section "Wettability and Contact Angle," a speckled wettability form of intermediate wetting mimics the relative permeability characteristics of strongly water-wet conditions [122].

Capillary Pressure Curves   By convention, oil-water capillary pressure, P_c, is defined as the pressure in the oil phase, p_o, minus the pressure in the water phase, p_w:

$$P_c = p_o - p_w \qquad (1.79)$$

Depending on wettability and history of displacement, capillary pressure can be positive or negative. Figure 1.59 presents the effect of wettability on capillary pressure as related by Killins, Nielsen, and Calhoun [176]. Drainage and imbibition curves can have similarities, but the capillary pressure values are positive for strongly water-wet and negative for strongly oil-wet conditions. In the intermediate wettability case shown

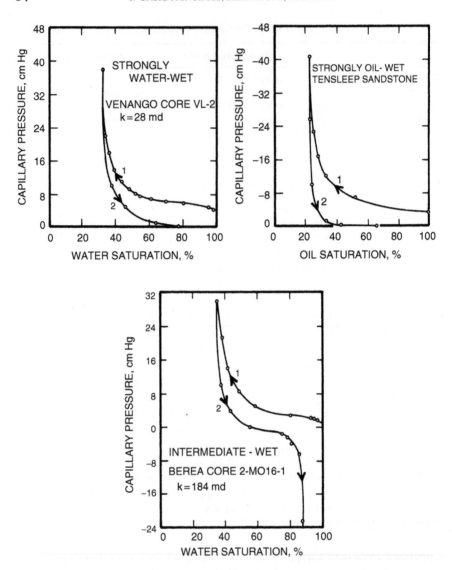

FIGURE 1.59   Effect of wettability on capillary pressure [176].

in Figure 1.59, the small positive value of threshold pressure during the drainage cycle suggests the sample was moderately water-wet [133]. After the drainage cycle, the sample spontaneously imbibed water until the capillary pressure was zero at a water saturation of 55%. Then, as water pressure was applied, the maximum water saturation of about 88% was achieved. As discussed previously, capillary pressure curves can be used as a criterion of wettability.

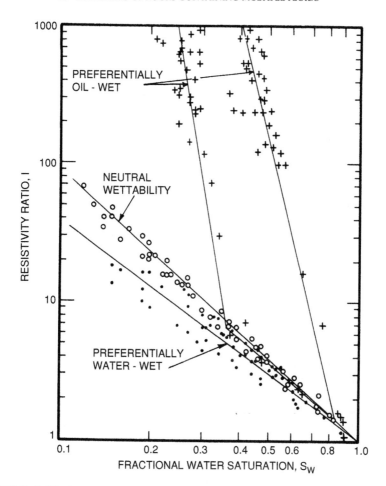

FIGURE 1.60 Resistivity-saturation relationships for different wettability carbonate cores [51].

Resistivity Factors and Saturation Exponents As shown previously in Table 1.7, Sweeney and Jennings [51] found that the formation resistivity factor changed when wettability was altered. However, the naphthenic acid they used to alter wettability may have also reduced porosity which could account for the increase in the saturation exponent in Equation 1.46. Other investigators [177, 178] have found no significant effect of wettability on formation factors. Because of the scarcity of data and the difficulty of altering wettability without affecting other properties, the effect of wettability on formation resistivity remains unclear.

As related by Mungan and Moore [178], three assumptions are made in the Archie saturation expression (Equation 1.68): all of the brine contributes to electrical flow: the saturation exponent, n, is constant for a given porous media; and only one resistivity is measured for a given saturation. Since the saturation exponent depends on the distribution of the conducting phase, which is dependent on wettability, the foregoing assumptions are valid only for strongly water-wet conditions. When wettability is altered, the differences in fluid distribution cause variations in the cross-sectional areas of conductive paths and in the tortuous path lengths. These variations affect resistivity, which results in different resistivity-saturation relationships such as were presented for carbonate cores by Sweeney and Jennings [51] in Figure 1.60. The saturation exponent, which is the slope of the lines, was about 1.6 for water-wet cores, about 1.9 for neutral-wet cores, and about 8 for oil-wet cores [51]. Similar data in sandstone cores were provided by Rust [177]; saturation exponents were about 1.7 and 13.5 for water-wet and oil-wet conditions respectively. These differences in oil-wet rocks most likely occur because of the isolation of trapped brine in dendritic fingers or dead-end pores which do not contribute to electrical conductivity.

Table 1.19, presenting data given by Mungan and Moore [178], shows that the effect of wettability on saturation exponent becomes more important at low brine saturations. Morgan and Pirson [179] conducted tests on fractionally wetted bead packs in which portions of the beads were water-wet and portions were treated so that they were mildly oil-wet. From their data, plotted in Figure 1.61, the saturation exponent increased as the extent of oil-wetting increased.

The foregoing data suggest that unless the reservoir is known to be water-wet, the saturation exponent should be measured with native-state (preferably) or restored cores. If the reservoir is oil-wet and clean cores

TABLE 1.19   Saturation Exponents in Teflon Cores Partially Saturated with Nonwetting Conducting Liquid

Air-NaCl solution		Oil-NaCl solution	
Brine Saturation % PV	n	Brine Saturation % PV	n
66.2	1.97	64.1	2.35
65.1	1.98	63.1	2.31
63.2	1.92	60.2	2.46
59.3	2.01	55.3	2.37
51.4	1.93	50.7	2.51
43.6	1.99	44.2	2.46
39.5	2.11	40.5	2.61
33.9	4.06	36.8	2.81
30.1	7.50	34.3	4.00
28.4	8.90	33.9	7.15
		31.0	9.00

From Reference 178.

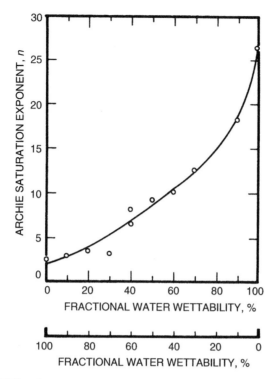

FIGURE 1.61 Influence of wettability on saturation exponent [179].

are used that may be water-wet, the saturation exponents that are obtained can lead to an underestimate of connate water saturation in the formation tested.

References

[1] Standing, M. B., "A Pressure-Volume-Temperature Correlation for Mixtures of California Oils and Gases," *Drill. & Prod. Prac.*, API (1947), pp. 275–287.
[2] Standing, M. B., *Volumetric and Phase Behavior of Oil Field Hydrocarbon Systems*, Reinhold Publishing Corp., New York (1952).
[3] Katz, D. L., et al., *Handbook of Natural Gas Engineering*, McGraw-Hill Book Co., Inc., New York (1959).
[4] Hocott, C. R., and Buckley, S. E., "Measurements of the Viscosities of Oils under Reservoir Conditions," *Trans.*, AIME, Vol. 142 (1941), pp. 131–136.
[5] Beal, C., "The Viscosity of Air, Water, Natural Gas, Crude Oil and Its Associated Gases at Oil Field Temperature and Pressures," *Trans.*, AIME, Vol. 165 (1946), pp. 94–115.
[6] Chew, J. N., and Connally, C. A., "A Viscosity Correlation for Gas-Saturated Crude Oils," *Trans.*, AIME, Vol. 216 (1959), pp. 23–25.
[7] Beggs, H. D., and Robinson, J. R., "Estimating the Viscosity of Crude Oil Systems," *J. Pet. Tech.* (Sept. 1975), pp. 1140–1141.

[8] Lohrenz, J., Bray, B. G., and Clark, C. R., "Calculating Viscosities of Reservoir Fluids from Their Compositions," *J. Pet. Tech.* (Oct. 1964), pp. 1171–1176; *Trans.*, AIME, Vol. 231.

[9] Houpeurt, A. H., and Thelliez, M. B., "Predicting the Viscosity of Hydrocarbon Liquid Phases from Their Composition," paper SPE 5057 presented at the SPE 49th Annual Fall Meeting, Houston, Oct. 6–9, 1974.

[10] Little, J. E., and Kennedy, H. T., "A Correlation of the Viscosity of Hydrocarbon Systems with Pressure, Temperature and Composition," *Soc. Pet. Eng. J.* (June 1968), pp. 157–162; *Trans.*, AIME, Vol. 243.

[11] Swindells, J. F., Coe, J. R., and Godfrey, T. B., "Absolute Viscosity of Water at 20°C," Res. Paper 2279, *J. Res. Nat'l Bur. Stnds.*, Vol. 48 (1952), pp. 1–31.

[12] *CRC Handbook of Chemistry and Physics*, 62nd edition, R. C. Weast (Ed.), CRC Press, Inc., Boca Raton, FL (1982).

[13] Matthews, C. S., and Russell, D. G., *Pressure Buildup and Flow Tests in Wells*, Monograph Series, SPE, Dallas (1967), p. 1.

[14] Pirson, S. J., *Oil Reservoir Engineering*, McGraw-Hill Book Co., Inc., New York (1958).

[15] Kirkbride, C. G., *Chemical Engineering Fundamentals*, McGraw-Hill Book Co., Inc., New York (1947), p. 61.

[16] Standing, M. B., and Katz, D. L., "Density of Natural Gases," *Trans.*, AIME, Vol. 146 (1942), pp. 140–149.

[17] Craft, B. C., and Hawkins, M. F., *Applied Petroleum Reservoir Engineering*, Prentice-Hall, Inc., Englewood Cliffs (1959).

[18] McCain, W. D., *The Properties of Petroleum Fluids*, Petroleum Publishing Co., Tulsa (1973).

[19] Amyx, J. W., Bass, D. M., Jr., and Whiting, R. L., *Petroleum Reservoir Engineering*, McGraw-Hill Book Co., Inc., New York (1960).

[20] Slider, H. C., *Practical Petroleum Engineering Methods*, Petroleum Publishing Co., Tulsa (1976).

[21] Taylor, L. B., personal communication.

[22] Dodson, C. R., and Standing, M. B., "Pressure-Volume-Temperature and Solubility Relations for Natural Gas-Water Mixtures," *Drill. & Prod. Prac.*, API (1944), pp. 173–179.

[23] Wyllie, M. R. J., *Petroleum Production Handbook*, T. C. Frick (Ed.), Vol. II, Reservoir Engineering, SPE, Dallas (1962).

[24] Burcik, E. J., *Properties of Petroleum Fluids*, Int'l Human Resources Dev. Corp., Boston (1979).

[25] Muskat, M., *Physical Principles of Oil Production*, McGraw-Hill Book Co., Inc., New York (1949).

[26] Ramey, H. J., Jr., "Rapid Methods for Estimating Reservoir Compressibilities," *J. Pet. Tech.* (April 1964), pp. 447–454.

[27] Trube, A. S., "Compressibility of Natural Gases," *Trans.*, AIME, Vol. 210 (1957), pp. 355–357.

[28] Trube, A. S., "Compressibility of Undersaturated Hydrocarbon Reservoir Fluids," *Trans.*, AIME, Vol. 210 (1957), pp. 341–344.

[29] "A Correlation for Water Viscosity," *Pet. Eng.* (July 1980), pp. 117–118.

[30] Standing, M. B., *Volumetric and Phase Behavior of Oil Field Hydrocarbon Systems*, SPE, Dallas (1977).

[31] Carr, N. L., Kobayashi, R., and Burrows, D. B., "Viscosity of Hydrocarbon Gases Under Pressure," *Trans.*, AIME, Vol. 201 (1954), pp. 264–272.

[32] Hollo, R., and Fifadara, H., *TI-59 Reservoir Engineering Manual*, PennWell Books, Tulsa (1980).

[33] Vazquez, M., and Beggs, H. D., "Correlations for Fluid Physical Property Prediction," *J. Pet. Tech.* (June 1980), pp. 968–970; paper SPE 6719 presented at the 52nd Annual Fall Tech. Conf. and Exhibition, Denver, Oct. 9–12, 1977.

[34] Meehan, D. N., "Improved Oil PVT Property Correlations," *Oil & Gas J.* (Oct. 27, 1980), pp. 64–71.

[35] Meehan, D. N., "Crude Oil Viscosity Correlation," *Oil & Gas J.* (Nov. 10, 1980), pp. 214–216.

[36] Meehan, D. N., and Vogel, E. L., *HP-41 Reservoir Engineering Manual*, PennWell Books, Tulsa (1982).

[37] Garb, F. A., *Waterflood Calculations for Hand-Held Computers*, Gulf Publishing Co., Houston (1982).

[38] Hollo, R., Homes, M., and Pais, V., *HP-41CV Reservoir Economics and Engineering Manual*, Gulf Publishing Co., Houston (1983).

[39] McCoy, R. L., *Microcomputer Programs for Petroleum Engineers, Vol 1. Reservoir Engineering and Formation Evaluation*, Gulf Pub. Co., Houston (1983).

[40] Sinha, M. K., and Padgett, L. R., *Reservoir Engineering Techniques Using Fortran*, Int'l Human Resources Dev. Corp., Boston (1985).

[41] Smith, H. I., "Estimating Flow Efficiency From Afterflow-Distorted Pressure Buildup Data," *J. Pet. Tech.* (June 1974), pp. 696–697.

[42] Archie, G. E., "The Electrical Resistivity Log as an Aid in Determining Some Reservoir Characteristics," *Trans.*, AIME (1942), pp. 54–61.

[43] Wyllie, M. R. J., and Rose, W. D., "Some Theoretical Considerations Related to the Quantitative Evaluation of the Physical Characteristics of Reservoir Rock from Electrical Log Data," *Trans.*, AIME, Vol. 189 (1950), pp. 105–118.

[44] Tixier, M. P., "Porosity Index in Limestone from Electrical Logs," *Oil & Gas J.* (Nov. 15, 1951), pp. 140–173.

[45] Winsauer, W. O., "Resistivity of Brine-Saturated Sands in Relation to Pore Geometry," *Bull.*, AAPG, Vol. 36, No. 2 (1952), pp. 253–277.

[46] Wyllie, M. R. J., and Gregory, A. R., "Formation Factors of Unconsolidated Porous Media: Influence of Particle Shape and Effect of Cementation," *Trans.*, AIME, Vol. 198 (1953), pp. 103–109.

[47] Cornell, D., and Katz, D. L., "Flow of Gases Through Consolidated Porous Media," *Ind. & Eng. Chem.*, Vol. 45 (Oct. 1953), pp. 2145–2152.

[48] Owen, J. D., "Well Logging Study-Quinduno Field, Roberts County, Texas," Symp. on Formation Evaluation, AIME (Oct. 1955).

[49] Hill, H. J., and Milburn, J. D., "Effect of Clay and Water Salinity on Electrochemical Behavior of Reservoir Rocks," *Trans.*, AIME, Vol. 207 (1956), pp. 65–72.

[50] Wyllie, M. R. J., and Gardner, G. H. F., "The Generalized Kozeny-Carman Equation, Part 2–A Novel Approach to Problems of Fluid Flow," *World Oil* (April 1958), pp. 210–228.

[51] Sweeney, S. A., and Jennings, H. Y., Jr., "Effect of Wettability on the Electrical Resistivity of Carbonate Rock from a Petroleum Reservoir," *J. Phys. Chem.*, Vol. 64 (1960), pp. 551–553.

[52] Carothers, J. E., "A Statistical Study of the Formation Factor Relation," *The Log Analyst* (Sept.–Oct. 1968), pp. 13–20.

[53] Porter, C. R., and Carothers, J. E., "Formation Factor-Porosity Relation Derived from Well Log Data," *The Log Analyst* (Jan.–Feb. 1971), pp. 16–26.

[54] Timur, A., Hemkins, W. B., and Worthington, A. W., "Porosity and Pressure Dependence of Formation Resistivity Factor for Sandstones," Proc. Cdn. Well Logging Soc., Fourth Formation Evaluation Symposium, Calgary, Alberta, May 9–12, 1972.

[55] Perez-Rosales, C., "On the Relationship Between Formation Resistivity Factor and Porosity," *Soc. Pet. Eng. J.* (Aug. 1982), pp. 531–536.

[56] Coates, G. R., and Dumanoir, J. L., "A New Approach to Log-Derived Permeability," *Trans.*, SPWLA 14th Annual Logging Symp. (May 6–9, 1973), pp. 1–28.

[57] Amyx, J. W., and Bass, D. M., Jr., "Properties of Reservoir Rocks," Chap. 23 in *Pet. Prod. Handbook*, T. C. Frick and R. W. Taylor (Eds.), SPE, Dallas, 2 (1962) 23/1–23/40.

[58] *Log Interpretation–Principles and Applications*, Schlumberger Educational Services, Houston (1972).

[59] Patnode, H. W., and Wyllie, M. R. J., "The Presence of Conductive Solids in Reservoir Rocks as a Factor in Electric Log Interpretation," *Trans.*, AIME, Vol. 189 (1950), pp. 47–52.

[60] Winn, R. H., "The Fundamentals of Quantitative Analysis of Electric Logs," *Proc.*, Symposium on Formation Evaluation (Oct. 1955), pp. 35–48.

[61] Atkins, E. R., and Smith, G. H., "The Significance of Particle Shape in Formation Resistivity Factor-Porosity Relationships," *J. Pet . Tech.* (March 1961); *Trans.*, AIME, Vol. 222, pp. 285–291.

[62] Koepf, E. H., "Core Handling-Core Analysis Methods," Chap. 3 in *Determination of Residual Oil Saturation*, Interstate Oil Compact Commission, Oklahoma City, Oklahoma (1978), pp. 36–71.

[63] Sanyal, S. K., Marsden, S. S., Jr., and Ramey, H. J., Jr., "The Effect of Temperature on Electrical Resistivity of Porous Media," *The Log Analyst* (March–April 1973), pp. 10–24.

[64] Hilchie, D. W., "The Effect of Pressure and Temperature on the Resistivity of Rocks," Ph.D. Dissertation, The University of Oklahoma (1964).

[65] Ransom, R. C., "A Contribution Toward a Better Understanding of the Modified Archie Formation Resistivity Factor Relationship," *The Log Analyst* (March–April 1984), pp. 7–12.

[66] Earlougher, R. C., Jr., *Advances in Well Test Analysis*, Monograph Series, SPE, Dallas (1977), Vol. 5.

[67] Fatt, I., "Pore Volume Compressibilities of Sandstone Reservoir Rock," *J. Pet. Tech.* (March 1958), pp. 64–66.

[68] Van der Knapp, W., "Nonlinear Behavior of Elastic Porous Media," *Trans.*, AIME, Vol. 216 (1959), pp. 179–187.

[69] Hall, H. N., "Compressibility of Reservoir Rocks," *Trans.*, AIME, Vol. 198 (1953), pp. 309–311.

[70] Carpenter, C. B., and Spencer, G. B., "Measurements of Compressibility of Consolidated Oil-Bearing Sandstones," RI 3540, USBM (Oct., 1940).

[71] Newman, G. H., "Pore-Volume Compressibility of Consolidated, Friable, and Unconsolidated Reservoir Rocks Under Hydrostatic Loading," *J. Pet. Tech.* (Feb. 1973), pp. 129–134.

[72] Martin, J. C., "Simplified Equations of Flow in Gas Drive Reservoirs and the Theoretical Foundation of Multiphase Pressure Buildup Analyses," *Trans.*, AIME, Vol. 216 (1959), pp. 309–311.

[73] Calhoun, J. C., Jr., *Fundamentals of Reservoir Engineering*, U. of Oklahoma Press, Norman (1976).

[74] Martin, M., Murray, G. H., and Gillingham, W. J., "Determination of the Potential Productivity of Oil-Bearing Formations by Resistivity Measurements," *Geophysics*, Vol. 3 (1938), pp. 258–272.

[75] Jakosky, J. J., and Hopper, R. H., "The Effect of Moisture on the Direct Current Resistivities of Oil Sands and Rocks," *Geophysics*, Vol. 2 (1937), pp. 33–55.

[76] Wyckoff, R. D., and Botset, H. G., "The Flow of Gas-Liquid Mixtures Through Unconsolidated Sands," *J. Applied Physics*, Vol. 7 (Sept. 1936), pp. 325–345.

[77] Leverett, M. C., "Flow of Oil-Water Mixtures Through Unconsolidated Sands," *Trans.*, AIME (1938), pp. 149–171.

[78] de Witte, L., "Relations Between Resistivities and Fluid Contents of Porous Rocks," *Oil & Gas J.* (Aug. 24, 1950), pp. 120–132.

[79] de Witte, A. J., "Saturation and Porosity From Electric Logs in Shaly Sands," *Oil & Gas J.* (March 4, 1957), pp. 89–93.

[80] Williams, M., "Estimation of Interstitial Water from the Electrical Log," *Trans.*, AIME, Vol. 189 (1950), pp. 295–308.

[81] Rust, C. F., "Electrical Resistivity Measurements on Reservoir Rock Samples by the Two-Electrode and Four-Electrode Methods," *Trans.*, AIME, Vol. 192 (1952), pp. 217–224.

[82] Wyllie, M. R. J., and Spangler, M. B., "Application of Electrical Resistivity Measurements to Problem of Fluid Flow in Porous Media," *Bull.*, AAPG, Vol. 36, No. 2 (1952), pp. 359–403.

[83] Morse, R., Terwilliger, P. L., and Yuster, S. T., "Relative Permeability Measurements on Small Core Samples," *Oil & Gas J.* (Aug. 23, 1947), pp. 109–125.

[84] Whiting, R. L., Guerrero, E. T., and Young, R. M., "Electrical Properties of Limestone Cores," *Oil & Gas J.* (July 27, 1953), pp. 309–313.

[85] Dunlap, H. F., et al., "The Relation Between Electrical Resistivity and Brine Saturation in Reservoir Rocks," *Trans.*, AIME, Vol. 186 (1949), pp. 259–264.

[86] Dorfman, M. H., "Discussion of Reservoir Description Using Well Logs," *J. Pet. Tech.* (Dec. 1984), pp. 2195–2196.

[87] Wilson, D. A., and Hensel, W. M. Jr., "The Cotton Valley Sandstones of East Texas: a Log-Core Study," *Trans.*, SPWLA, 23rd Annual Logging Symposium, Paper R (July 6–9, 1982), pp. 1–27.

[88] Hilchie, D. W., "Author's Reply to Discussion of Reservoir Description Using Well Logs," *J. Pet. Tech.* (Dec. 1984), p. 2196.

[89] Hocott, C. R., "Interfacial Tension Between Water and Oil Under Reservoir Conditions," *Pet. Tech.* (Nov. 1938), pp. 184–190.

[90] Katz, D. L., Monroe, R. R., and Trainer, R. P., "Surface Tension of Crude Oils Containing Dissolved Gases," *Pet. Tech.* (Sept. 1943), pp. 1–10.

[91] Swartz, C. A., "The Variation in the Surface Tension of Gas-Saturated Petroleum with Pressure of Saturation," *Physics*, Vol. 1 (1931), pp. 245–253.

[92] Livingston, H. K., "Surface and Interfacial Tensions of Oil-Water Systems in Texas Oil Sands," *Pet. Tech.* (Nov. 1938), pp. 1–13.

[93] Hjelmeland, O. S., and Larrondo, L. E., "Investigation of the Effects of Temperature, Pressure, and Crude Oil Composition on Interfacial Properties," *SPE Reservoir Engineering* (July 1986), pp. 321–328.

[94] Defay, R., Prigogine, I., Bellemans, A., and Everett, D. H., *Surface Tension and Adsorption*, Longmans, London (1966).

[95] Benner, F. C., Riches, W. W., and Bartell, F. E., "Nature and Importance of Surface Forces in Production of Petroleum," *Drill & Prod. Prac.*, API (1938), pp. 442–448.

[96] Benner, F. C., and Bartell, F. E., "The Effect of Polar Impurities Upon Capillary and Surface Phenomena in Petroleum Production," *Drill & Prod. Prac.*, API (1942), pp. 341–348.

[97] Benner, F. C., Dodd, C. G., and Bartell, F. E., "Evaluation of Effective Displacement Pressures for Petroleum Oil-Water Silica Systems," *Drill & Prod. Prac.*, API (1942), pp. 169–177.

[98] Nutting, P. G., "Some Physical and Chemical Properties of Reservoir Rocks Bearing on the Accumulation and Discharge of Oil," *Problems in Petroleum Geology*, AAPG (1934).

[99] Schilthuis, R. J., "Connate Water in Oil and Gas Sands," *Trans.*, AIME (1938), pp. 199–214.

[100] Leverett, M. C., "Capillary Behavior in Porous Solids," *Trans.*, AIME, Vol. 142 (1941), pp. 159–172.

[101] Bartell, F. E., and Miller, F. L., "Degree of Wetting of Silica by Crude Petroleum Oils," *Ind. Eng. Chem.*, Vol. 20, No. 2 (1928), pp. 738–742.

[102] Slobod, R. L., and Blum, H. A., "Method for Determining Wettability of Reservoir Rocks," *Trans.*, AIME, Vol. 195 (1952), pp. 1–4.

[103] Taber, J. J., personal communication.

[104] Wagner, O. R., and Leach, R. O., "Improving Oil Displacement Efficiency by Wettability Adjustment," *Trans.*, AIME, Vol. 216 (1959), pp. 65–72.

[105] Denekas, M. O., Mattax, C. C., and Davis, G. T., "Effects of Crude Oil Components on Rock Wettability," *Trans.*, AIME, Vol. 216 (1959), pp. 330–333.

[106] Moore, T. F., and Slobod, R. L., "The Effect of Viscosity and Capillarity on the Displacement of Oil by Water," *Prod. Monthly* (Aug. 1956), pp. 20–30.

[107] Bobek, J. E., Mattax, C. C., and Denekas, M. O., "Reservoir Rock Wettability—Its Significance and Evaluation," *Trans.*, AIME, Vol. 213 (1958), pp. 155–160.

[108] Amott, E., "Observations Relating to the Wettability of Porous Rock," *Trans.*, AIME, Vol. 216 (1959), pp. 156–162.

[109] Gatenby, W. A., and Marsden, S. S., "Some Wettability Characteristics of Synthetic Porous Media," *Prod. Monthly* (Nov. 1957), pp. 5–12.

[110] Donaldson, E. C., Thomas, R. D., and Lorenz, P. B., "Wettability Determination and Its Effect on Recovery Efficiency," *Soc. Pet. Eng. J* (March 1969), pp. 13–20.

[111] Donaldson, E. C., et al., "Equipment and Procedures for Fluid Flow and Wettability Tests of Geological Materials," U.S. Dept. of Energy, Bartlesville, Report DOE/BETC/IC-79/5, May 1980.

[112] Donaldson, E. C., "Oil-Water-Rock Wettability Measurement," *Proc.*, Symposium of Chemistry of Enhanced Oil Recovery, Div. Pet. Chem., Am. Chem. Soc., March 29–April 3, 1981, pp. 110–122.

[113] Raza, S. H., Treiber, L. E., and Archer, D. L., "Wettability of Reservoir Rocks and Its Evaluation," *Prod. Monthly*, Vol. 33, No. 4 (April 1968), pp. 2–7.

[114] Treiber, L. E., Archer, D. L., and Owens, W. W., "A Laboratory Evaluation of the Wettability of Fifty Oil-Producing Reservoirs," *Soc. Pet. Eng. J.* (Dec. 1972), pp. 531–540.

[115] Morrow, N. R., "Capillary Pressure Correlation for Uniformly Wetted Porous Media," *J. Can. Pet. Tech.*, Vol. 15 (1976), pp. 49–69.

[116] Chilingar, G. V., and Yen, T. F., "Some Notes on Wettability and Relative Permeabilities of Carbonate Reservoir Rocks, II," *Energy Sources*, Vol. 7, No. 1 (1983), pp. 67–75.

[117] Brown, R. J. S., and Fatt, I., "Measurements of Fractional Wettability of Oil Field Rocks by the Nuclear Magnetic Relaxation Method," *Trans.*, AIME, Vol. 207 (1956), pp. 262–264.

[118] Fatt, I., and Klikoff, W. A., "Effect of Fractional Wettability on Multiphase Flow through Porous Media," *Trans.*, AIME, Vol. 216 (1959), pp. 426–432.

[119] Holbrook, O. C., and Bernard, G. C., "Determination of Wettability by Dye Adsorption," *Trans.*, AIME, Vol. 213 (1958), pp. 261–264.

[120] Iwankow, E. N., "A Correlation of Interstitial Water Saturation and Heterogenous Wettability," *Prod Monthly* (Oct. 1960), pp. 18–26.

[121] Salathiel, R. A., "Oil Recovery by Surface Film Drainage in Mixed-Wettability Rocks," *Trans.*, AIME (1973), pp. 1216–1224.

[122] Morrow, N. R., Lim, H. T., and Ward, J. S., "Effect of Crude Oil Induced Wettability Changes on Oil Recovery," *SPE Formation Evaluation* (Feb. 1986), pp. 89–103.

[123] Rathmell, J. J., Braun, P. H., and Perkins, T. K., "Reservoir Waterflood Residual Oil Saturation from Laboratory Tests," *J. Pet. Tech.* (Feb. 1973), pp. 175–185.

[124] Sharma, M. M., and Wunderlich, R. W., "The Alteration of Rock Properties Due to Interactions With Drilling Fluid Components," paper SPE 14302 presented at the SPE 1985 Annual Technical Conference & Exhibition, Las Vegas, Sept. 22–25.

[125] Richardson, J. G., Perkins, F. M., Jr., and Osoba, J. S., "Differences in Behavior of Fresh and Aged East Texas Woodbine Cores," *Trans.*, AIME, Vol. 204 (1955), pp. 86–91.

[126] Mungan, N., "Certain Wettability Effects in Laboratory Waterfloods," *J. Pet. Tech.* (Feb. 1966), pp. 247–252.

[127] Rose, W. R., and Bruce, W. A., "Evaluation of Capillary Character in Petroleum Reservoir Rock," *Trans.*, AIME, Vol. 186 (1949), pp. 127–133.

[128] Klinkenberg, L. J., "The Permeability of Porous Media to Liquids and Gases," *Drill. & Prod. Prac.*, API (1941), pp. 200–213.

[129] Jones, F. O., and Owens, W. W., "A Laboratory Study of Low-Permeability Gas Sands," *J. Pet. Tech.* (Sept. 1980), pp. 1631–1640.

[130] Wei, K. K., Morrow, N. R., and Brower, K. R., "The Effect of Fluid, Confining Pressure, and Temperature on Absolute Permeabilities of Low-Permeability Sandstones," *SPE Formation Evaluation* (Aug. 1986), pp. 413–423.

[131] Fatt, I., "The Effect of Overburden Pressure on Relative Permeability," *Trans.*, AIME (1953), pp. 325–326.

[132] Bruce, W. A., and Welge, H. J., "The Restored-State Method for Determination of Oil in Place and Connate Water," *Drill. and Prod. Prac.*, API (1947), pp. 166–174.

[133] Craig, F. F., Jr., "The Reservoir Engineering Aspects of Waterflooding," Monograph Series, SPE, Dallas, Vol. 3 (1971).

[134] Muskat, M., et al., "Flow of Gas-Liquid Mixtures through Sands," *Trans.*, AIME (1937), pp. 69–96.

[135] Hassler, G. L., Rice, R. R., and Leeman, E. H., "Investigations on the Recovery of Oil from Sandstones by Gas Drive," *Trans.*, AIME (1936), pp. 116–137.

[136] Botset, H. G., "Flow of Gas-Liquid Mixtures through Consolidated Sand," *Trans.*, AIME Vol. 136 (1940), pp. 91–105.

[137] Leverett, M. C., and Lewis, W. B., "Steady Flow of Gas-Oil Water Mixtures through Unconsolidated Sands," *Trans.*, AIME, Vol. 142 (1941), pp. 107–116.

[138] Krutter, H., and Day, R. J., "Air Drive Experiments on Long Horizontal Consolidated Cores," *Pet. Tech.*, T. P. 1627 (Nov. 1943).

[139] Osoba, J. S., et al., "Laboratory Measurements of Relative Permeability," *Trans.*, AIME, Vol. 192 (1951), pp. 47–56.

[140] Caudle, B. H., Slobod, R. L., and Brownscombe, E. R., "Further Developments in the Laboratory Determination of Relative Permeability," *Trans.*, AIME, Vol. 192 (1951), pp. 145–150.

[141] Geffen, J. M., et al., "Experimental Investigation of Factors Affecting Laboratory Relative Permeability Measurements," *Trans.*, AIME, Vol. 192 (1951), pp. 99–110.

[142] Richardson, J. G., et al., "Laboratory Determination of Relative Permeability," *Trans.*, AIME, Vol. 195 (1952), pp. 187–196.

[143] Owens, W. W., Parrish, D. R., and Lamoreaux, W. E., "An Evaluation of a Gas-Drive Method for Determining Relative Permeability Relationships," *Trans.*, AIME, Vol. 207 (1956), pp. 275–280.

[144] Hassler, G. L., "Method and Apparatus for Permeability Measurements," U. S. Patent No. 2,345,935.

[145] Brownscombe, E. R., Slobod, R. L., and Caudle, B. H., "Relative Permeability of Cores Desaturated by Capillary Pressure Method," *Drill & Prod. Prac.*, API (1949), pp. 302–315.

[146] Gates, J. I., and Tempelaar-Lietz, W., "Relative Permeabilities of California Cores by the Capillary Pressure Method," *Drill & Prod. Prac.*, API (1950), pp. 285–302.

[147] Leas, W. J., Jenks, L. H., and Russell, C. D., "Relative Permeability to Gas," *Trans.*, AIME, Vol. 189 (1950), pp. 65–72.

[148] Rapoport, L. A., and Leas, W. J., "Relative Permeability to Liquid in Liquid-Gas Systems," *Trans.*, AIME, Vol. 192 (1951), pp. 83–98.

[149] Fatt, I., and Dykstra, H., "Relative Permeability Studies," *Trans.*, AIME, Vol. 192 (1951), pp. 249–256.

[150] Corey, A. T., "The Interrelation Between Gas and Oil Relative Permeabilities," *Prod. Monthly*, Vol. 19 (Nov. 1954), pp. 38–41.

[151] Loomis, A. G., and Crowell, D. C., "Relative Permeability Studies: Gas-Oil and Water-Oil Systems," Bull. 599, U. S. Bureau of Mines, Washington, 1962.

[152] Buckley, S. E., and Leverett, M. C., "Mechanism of Fluid Displacement in Sands," *Trans.*, AIME, Vol. 146 (1942), pp. 107–116.

[153] Welge, H. J., "A Simplified Method for Computing Oil Recovery by Gas or Water Drive," *Trans.*, AIME, Vol. 195 (1952), pp. 91–98.

[154] Johnson, E. F., Bossler, D. P., and Naumann, V. O., "Calculation of Relative Permeability from Displacement Experiments," *Trans.*, AIME, Vol. 216 (1959), pp. 370–372.

[155] Higgins, R. V., "Application of Buckley-Leverett Techniques in Oil-Reservoir Analysis," Bureau of Mines Report of Investigations 5568 (1960).

[156] Guerrero, E. T., and Stewart, F. M., "How to Obtain a k_w/k_o Curve from Laboratory Unsteady-State Flow Measurements," *Oil & Gas J.* (Feb. 1, 1960), pp. 96–100.

[157] Guerrero, E. T., and Stewart, F. M., "How to Obtain and Compare k_w/k_o Curves from Steady-State and Laboratory Unsteady-State Flow Measurements," *Oil & Gas J.* (Feb. 22, 1960), pp. 104–106.

[158] Jones, S. C., and Rozelle, W. O., "Graphical Techniques for Determining Relative Permeability from Displacement Experiments," *Trans.*, AIME (1978), pp. 807–817.

[159] Corey, A. T., et al., "Three-Phase Relative Permeability," *J. Pet. Tech.* (Nov. 1956), pp. 63–65.

[160] Schneider, F. N., and Owens, W. W., "Sandstone and Carbonate Two- and Three-Phase Relative Permeability Characteristics," *Soc. Pet. Eng. J.* (March 1970), pp. 75–84.

[161] Honarpour, M., Koederitz, L. F., and Harvey, A. H., "Empirical Equations for Estimating Two-Phase Relative Permeability in Consolidated Rock," *J. Pet. Tech.* (Dec. 1982), pp. 2905–2908.

[162] Mungan, N., "Interfacial Effects in Immiscible Liquid-Liquid Displacement in Porous Media," *Soc. Pet. Eng. J.* (Sept. 1966), pp. 247–253.

[163] Lefebvre du Prey, E. J., "Factors Affecting Liquid-Liquid Relative Permeabilities of a Consolidated Porous Medium," *Soc. Pet. Eng. J.* (Feb. 1973), pp. 39–47.

[164] Miller, M. A., and Ramey, H. J., Jr., "Effect of Temperature on Oil/Water Relative Permeabilities of Unconsolidated and Consolidated Sands," *Soc. Pet. Eng. J.* (Dec. 1985), pp. 945–953.

[165] Honarpour, M., DeGroat, C., and Manjnath, A., "How Temperature Affects Relative Permeability Measurement," *World Oil* (May 1986), pp. 116–126.

[166] Kinney, P. T., and Nielsen, R. F., "Wettability in Oil Recovery," *World Oil*, Vol. 132, No. 4 (March 1951), pp. 145–154.

[167] Newcombe, J., McGhee, J., and Rzasa, M. J., "Wettability versus Displacement in Water Flooding in Unconsolidated Sand Columns," *Trans.*, AIME, Vol. 204 (1955), pp. 227–232.

[168] Coley, F. N., Marsden, S. S., and Calhoun, J. C., Jr., "A Study to the Effect of Wettability on the Behavior of Fluids in Synthetic Porous Media," *Prod. Monthly*, Vol. 20, No. 8 (June 1956), pp. 29–45.

[169] Jennings, H. Y., Jr., "Surface Properties of Natural and Synthetic Porous Media," *Prod. Monthly*, Vol. 21, No. 5 (March 1957), pp. 20–24.

[170] Uren, L. D., and Fahmy, E. H., "Factors Influencing the Recovery of Petroleum from Unconsolidated Sands by Water-Flooding," *Trans.*, AIME, Vol. 77 (1927), pp. 318–335.

[171] Kennedy, H. T., Burja, E. O., and Boykin, R. S., "An Investigation of the Effects of Wettability on the Recovery of Oil by Water Flooding," *J. Phys. Chem.*, Vol. 59 (1955), pp. 867–869.

[172] Morrow, N. R., "Interplay of Capillary, Viscous and Buoyancy Forces in the Mobilization of Residual Oil," *J. Can. Pet. Tech.* (July–Sept. 1979), pp. 3546.

[173] Melrose, J. C., and Brandner, C. F., "Role of Capillary Forces in Determining Microscopic Displacement Efficiency for Oil Recovery by Waterflooding," *J. Can. Pet. Tech.*, Vol. 13 (1974), pp. 54–62.

[174] Owens, W. W., and Archer, D. L., "The Effect of Rock Wettability on Oil Water Relative Permeability Relationships," *J. Pet. Tech.* (July 1971), pp. 873–878.

[175] McCaffery, F. G., and Bennion, D. W., "The Effect of Wettability on Two-Phase Relative Permeabilities," *J. Can. Pet. Tech.*, Vol. 13 (1974), pp. 42–53.

[176] Killins, C. R., Nielsen, R. F., and Calhoun, J. C., Jr., "Capillary Desaturation and Imbibition in Rocks," *Prod. Monthly*, Vol. 18, No. 2 (Feb. 1953), pp. 30–39.

[177] Rust, C. F., "A Laboratory Study of Wettability Effects on Basic Core Parameters," paper SPE 986G, presented at the SPE Venezuelan Second Annual Meeting, Caracas, Venezuela, Nov. 6–9, 1957.

[178] Mungan, N., and Moore, E. J., "Certain Wettability Effects on Electrical Resistivity in Porous Media," *J. Can. Pet. Tech.* (Jan.–March 1968), pp. 20–25.

[179] Morgan, W. B., and Pirson, S. J., "The Effect of Fractional Wettability on the Archie Saturation Exponent," *Trans.*, SPWLA, Fifth Annual Logging Symposium, Midland, Texas, May 13–15, 1964.

Formation Evaluation

Formation evaluation, as applied to petroleum reservoirs, consists of the quantitative and qualitative interpretation of formation cores, geophysical well logs, mud logs, flow tests, pressure tests, and samples of reservoir fluids. The goal of the interpretation is to provide information concerning reservoir lithology, fluid content, storage capacity, and producibility of oil or gas reservoirs. The final analysis includes an economic evaluation of whether to complete an oil or gas well and, once completed, an ongoing analysis of how to produce the well most effectively. These interpretations and analyses are affected by geological complexity of the reservoir, rock quality, reservoir heterogeneity, and, from a logistical standpoint, the areal extent and location of the project of interest. In the early stages of development, the purpose of formation evaluation is to define reservoir thickness and areal extent, reservoir quality, reservoir fluid properties, and ranges of rock properties. The key rock properties are porosity, permeability, oil, gas, and water saturations. Because of space limitations and the importance of these properties, methods of measuring porosity, permeability, and fluid saturations will be emphasized.

2.1 CORING AND CORE ANALYSIS

Routine or conventional core analyses refer to common procedures that provide information on porosity, permeability, resident fluids, lithology, and texture of petroleum reservoirs. Table 2.1 lists the types of analyses that are obtained and how the results of each analysis are used. Specialized core analyses, such as are listed in Table 2.2, are done less often, but are important for specific applications. Routine core analyses can be performed on whole cores or on small plugs that are cut from a larger core. With the exception of petrographic analyses (thin sections, x-ray; scanning electron microscopy, etc.), special core analyses are normally done with core plugs. After a well is drilled and logs are available to identify zones of interest, very small portions of the reservoir can be obtained with percussion sidewall or sidewall drilled cores. Sidewall cores are less expensive and are valuable for petrographic analyses, but are generally not suitable for special core analyses.

The subject of coring and core analysis was summarized in a series of articles [2–10]. An overview article [11] described how core analyses can aid reservoir description. A handbook [12] is available that describes procedures and tools for conventional coring as well as methods for routine core analysis. Procedures for routine core analysis and methods of preserving cores have been recommended by the American Petroleum Institute [13]. Some of the information available in these sources will be highlighted.

2.1.1 Coring

Well coring refers to the process of obtaining representative samples of the productive formation in order to conduct a variety of laboratory testing. Various techniques are used to obtain core samples: conventional

TABLE 2.1 Routine Core Analysis Tests

Type of analysis	Use of results
Porosity	A factor in volume and storage determinations.
Permeability—horizontal and vertical	Defines flow capacity, crossflow, gas and water coning and relative profile capacity of different zones, pay and nonpay zones.
Saturations	Defines presence of hydrocarbons, probable fluid recovery by test, type of recovery, fluid contacts, completion interval.
Lithology	Rock type, fractures, vugs, laminations, shale content used in log interpretation, recovery forecasts, capacity estimates.
Core-gamma ray log	Relates core and log depth.
Grain density	Used in log interpretation and lithology.

From Reference 1.

TABLE 2.2   Special Core Analysis Tests

Type of test	Use of results
Capillary pressure	Defines irreducible fluid content, contacts.
Rock compressibility	Volume change caused by pressure change.
Permeability and porosity vs. pressure	Corrects to reservoir conditions.
Petrographic studies	
mineral	Used in log interpretation.
diagenesis	Origin of oil and source bed studies.
clay identification	Origin of oil and log analysis.
sieve analysis	Selection of screens, sand grain size.
Wettability	Used in capillary pressure interpretation and recovery analysis-relative permeability.
Electrical	
formation factor	Used in log interpretation.
resistivity index	
Acoustic velocity	Log and seismic interpretation.
Visual inspection	Rock description and geological study.
Thin sections, slabs	
Air, water, and other liquid permeability	Evaluates completion, workover, fracture and injection fluids; often combined with flood-pot test.
Flood-pot test and waterflood evaluation	Results in values for irreducible saturations, values for final recovery with special recovery fluids such as surfactants, water, and polymers.
Relative permeability	
gas-oil	Relative permeability is used to obtain values for effective
gas-water	permeability to each fluid when two or more fluids flow
water-oil	simultaneously; relative permeability enables the calcu-
oil-special fluids	lation of recovery versus saturation and time while val-
thermal	ues from flood-pot test give only end-point results.

From Reference 1.

diamond-bit coring, rubbersleeve coring, pressure coring, sidewall coring, and recovery of cuttings generated from the drilling operation. Conventional coring is normally done in competent formations to obtain full-diameter cores. Rubber sleeve-coring improves core recovery in softer formations. Pressure coring, although relatively expensive, is used to obtain cores that have not lost any fluids during lifting of the core to the surface.

A common problem with all of these techniques is to decide when to core. In many instances, cores from the interval of interest are not obtained because of abrupt stratigraphic changes. A second problem is that, typically, nonproductive intervals of the desired strata are obtained. These intervals did not initially contain a significant amount of hydrocarbon.

2.1.2 Core Preservation

The importance of not altering wettability with drilling mud filtrate has been discussed in Chapter 1 in the section entitled "Wettability and Contact Angle." Preventing wettability changes in core material, after it has been recovered at the surface, can be equally important so that subsequent laboratory measurements are representative of formation conditions.

Cores obtained with drilling muds that minimize wettability alteration, and that are protected at the well-site to prevent evaporation or oxidation, are called preserved cores. They are also referred to as fresh cores or native-state cores. Cores that are cleaned with solvents and resaturated with reservoir fluids are called restored-state cores or extracted cores. The restoring process is often performed on nonpreserved or weathered cores, but the same technique could apply to cores that had been preserved.

Two methods of preserving conventional cores, immediately after they have been removed from the core barrel, will prevent changes in wettability for several months. One method consists of immersing the core in deoxygenated formation brine or suitable synthetic brine (i.e., drilling mud filtrate) and keeping the samples in suitable containers that can be sealed to prevent leakage and the entrance of oxygen. In the second method, the cores are wrapped in Saran or polyethylene film and aluminum foil and then coated with wax or strippable plastic. The second method is preferred for cores that will be used for laboratory determination of residual oil content, but the first method may be preferred for laboratory displacement tests. Plastic bags are often recommended for short-term (2–4 days) storage of core samples. However, this method will not ensure unaltered rock wettability. Air-tight metal cans are not recommended because of the possibility of rust formation and potential leakage.

Cores taken with a pressure core barrel are often frozen at the well-site for transportation to the laboratory. (Cores are left in the inner core barrel.) Normally, the inner barrel containing the cores is cut into lengths convenient for transport. Because of the complexity of the operation, the pressure core barrel is not used as extensively as the conventional core barrel. An alternate procedure involves bleeding off the pressure in the core and core barrel while the produced liquids are collected and measured. Analysis of the depressured core is done by conventional techniques. Fluids collected from the barrel during depressuring are proportionately added to the volumes of liquid determined from core analysis. In this manner a reconstructed reservoir core saturation is provided.

2.1.3 Core Preparation

Depending on the type of core testing to be done, core samples may be tested as received in the laboratory or they may be cleaned to remove resident fluids prior to analysis. Details for cutting, cleaning, and preparing core plugs can be found in API RP-40: Recommended Practice for Core-Analysis Procedure [13], available from API Production Department, 211 North Ervay, Suite 1700, Dallas, TX 75201.

2.1.4 Core Analysis

Conventional core analysis procedures are described in detail in API RP-40 and elsewhere [12]. A good discussion on core analysis procedures is in the textbook written by Amyx, Bass, and Whiting [49].

2.1.4.1 Porosity

A number of methods [13] are suitable for measuring porosity of core samples. In almost all the methods, the sample is cleaned by solvent extraction and dried to remove liquid. Porosity can be determined by saturating the dry core with brine and measuring the weight increase after saturation. Another common method includes compressing a known volume of gas (usually helium) at a known pressure into a core that was originally at atmospheric pressure. Several other techniques have been used; one of the more common methods is the mercury porosimeter in which pressure on the core plug is reduced and the volume of the expanded air or gas is measured accurately. A summation of fluids technique, which measures and sums the oil, gas and water volumes in a freshly recovered reservoir core sample, is often used for plugs or sidewall samples of non-vuggy consolidated rocks that contain minimum amounts of clay [9].

Equations commonly used for calculation of porosity by gas expansion or compression include:

$$\phi = \left(\frac{V_p}{V_b}\right) 100 \qquad (2.1)$$

$$\phi = \left(\frac{V_b - V_{gr}}{V_b}\right) 100 \qquad (2.2)$$

where ϕ = porosity expressed as a percent
V_p = pore volume
V_b = bulk volume
V_{gr} = grain volume

All volumes should be in consistent units, commonly cm^3. If pore volume is measured directly in cores that contain vugs (such as some carbonates), Equation 2.1 may give erroneously high porosity because the bulk volume may be erroneously low [9]. If bulk volume of vuggy cores is determined by submerging the core in mercury or water, Equation 2.2 may yield erroneously low porosity [9]. Thus valid porosity values can only be obtained if bulk volume and grain volume measurements are accurate.

2.1.4.2 Permeability

The permeability of core plugs is determined by flowing a fluid (air gas, or water) through a core sample of known dimensions. If the absolute permeability is to be determined, the core plug is cleaned so that permeability is measured at 100% of the saturating fluid. Methods of measuring permeability of core plugs are described in API RP-27: Recommended Practice for Determining permeability of Porous Media [14]. Equation 1.36 can be used to calculate permeability of core plugs.

2.1.4.3 Fluid Saturations

Coring procedures usually alter the fluid content of the reservoir rock during the coring process. Drilling fluid is jetted against the formation rock ahead of the coring bit and the core surface as it enters the core barrel; as a result of this flushing action by the drilling mud filtrate, most free gas and a portion of the liquid are displaced from the core. When water base drilling fluid is used, the mud filtrate may displace oil until a condition of residual oil saturation is obtained. Also, this flushing action may result in the fluid content of the core being predominately that of the drilling fluid. When oil base drilling fluid is used, the core sample that is obtained may be driven to an irreducible water saturation.

2.1.4.4 Factors Affecting Oil Displaced During Coring

During the coring operation, it is important to avoid extreme flushing conditions that could cause mobilization of residual oil [15]. Some of the variables that control the amount of oil flushed from a core by mud filtrate are: borehole-to-formation differential pressure (overbalance), coring penetration rate, core diameter, type of drill bit, drilling mud composition (including particle size distribution), depth of invasion of mud particles into the core, rate of filtrate production (both spurt loss and total fluid loss), interfacial tension of mud filtrate, permeability of the formation (both horizontal and vertical), and nature of the reservoir (uniformity, texture, etc.). In one type of system investigated in the laboratory [16], the amount of oil stripped from cores varied directly with the overbalance pressure, filtration production rate, core diameter and core permeability; it varied inversely with penetration rate. In that system, the overbalance pressure exerted more influence than the other factors. When large pressure gradients exist near the core bit, unintentional displacement of residual oil may occur in coring operations. In this region close to the bit, high velocities caused by this high pressure may mobilize some of the residual oil. Drilling mud composition can affect subsequent laboratory oil displacement tests in core samples by: changing wettability of the reservoir rock, altering interfacial tension of the mud filtrate, being penetrated by mud particles into the zone of interest, and yielding undesirable fluid loss properties. Since fluids with lower interfacial tension contribute to additional oil recovery, whenever possible, the use of mud additives that lower interfacial tension should be avoided. Greater amounts of residual oil are displaced from cores as the filtrate production rate is increased. High API filter loss or smaller core diameters will generally lead to larger amounts of flushing, but a key factor in the amount of mobilized residual oil is the spurt loss (the rapid loss to the formation that occurs before an effective filter cake is formed). As stated previously, uniformity of the formation being cored will influence the amount of oil that will be displaced. Identical drilling conditions may yield varying results with changes in lithology or texture of the reservoir. In particular, drastic differences may be observed in reservoirs that contain both sandstone and carbonate oil-bearing strata.

2.1.4.5 Factors Affecting Oil Saturation Changes During Recovery of Cores

Surface oil saturations should be adjusted to compensate for shrinkage and bleedings [53]. Shrinkage is the term applied to the oil volume decrease caused by a temperature change or by a drop in pressure which causes dissolved gases to escape from solution. Shrinkage of reservoir fluids is measured in the laboratory by differentially liberating the samples at reservoir temperature. The formation volume factors are used to adjust surface oil volume back to reservoir temperature and pressure. Gases coming out of solution can cause some oil to flow out of the core even though it may have been flushed to residual oil by mud filtrate. Bleeding is the term applied to this decrease in oil saturation as the core is brought to the surface. Calculations have been proposed [53] to account for shrinkage and bleeding.

2.1.4.6 Measurement of Fluid Saturations

There are two primary methods of determining fluid content of cores; these methods are discussed in API RP-40: Recommended Practice for Core-Analysis Procedure [13]. In the retort or vacuum distillation method, a fluid content still is used to heat and vaporize the liquids under controlled conditions of temperature and pressure. Prior to testing, the gas space in the core is displaced with water. The fluids produced from the still are condensed and measured, and the fluid saturations are calculated. Normally the percent oil and water are subtracted from 100% to obtain the gas saturation; however, considerable error may be inherent in this assumption. The second common method is the distillation-extraction method in which water in the core is distilled, condensed, and accumulated in a calibrated receiving tube. Oil in the core is removed by solvent extraction and the oil saturation is calculated from the weight loss data and the water content data.

Conventional core samples have oil content determined by atmospheric distillation. The oil distilled from a sample is collected in a calibrated receiving tube where the volume is measured. Temperatures up to 1,200°F (about 650°C) are used to distill the oil from the sample which causes some coking and cracking of the oil and the loss of a small portion of the oil. An empirically derived correction is applied to the observed volume to compensate for the loss. Calibration tests are made on each type of oil.

Whole core samples have oil content determined by vacuum distillation. This technique is used to remove oil from the sample without destroying the minerals of the sample. A maximum temperature of 450°F is used. The oil distilled from the sample is collected in a calibrated receiving tube which is immersed in a cold bath of alcohol and dry ice at about −75°C. This prevents the oil from being drawn into the vacuum system. As in the atmospheric distillation method, corrections must be applied to the measured volumes.

The oil content (V_o) divided by the pore volume (V_p) yields the oil saturation (S_o) of a sample in percent of pore space:

$$S_o = \left(\frac{V_o}{V_p}\right) 100 \qquad (2.3)$$

Two sources of error are inherent in the retort method. At the high temperatures employed, water of crystallization within the rock is driven off which causes the water saturation to appear to be higher than the actual value. Another error results from the cracking of the oil and subsequent deposition of coke within the pore structure. Thus, a calibration curve should be prepared on various gravity crudes to compensate for the oil lost from the cracking and coking. Both of the above errors will result in a measured oil saturation that is lower than the actual saturation in the rock. Another possible source of error is the liberation of carbon dioxide from carbonate material in the core at elevated temperatures; this would cause a weight loss that can be interpreted as a change in saturation. The solvent extraction method has the disadvantage in that it is an indirect method since only the water removed from the core is measured. However the extraction method has the advantage that the core is usually not damaged and can be used for subsequent tests.

2.1.4.7 Grain Density and Core Description

Grain density and lithologic descriptions are often provided in data for routine care analysis. Grain density depends on the lithology and composition of the reservoir of interest. Densities of some common minerals found in reservoir rocks are listed in Table 2.3 [50].

2.1.4.8 Results of Core Analyses from Various Reservoirs

Typical core analyses [51] of different formations from various states and regions of the U.S. are listed in. Table 2.4a to 2.4i. In addition to ranges in permeability, porosity, oil saturation, connate water saturation, the depth and thickness of the productive intervals are given.

2.1.4.9 Special Core Analysis Tests

Special core analysis testing is done when specifically required. Visual inspection and some petrographic studies are frequently done. For sandstones and conglomerates, particle size is often obtained by disaggregating

TABLE 2.3 Densities of Common Minerals in Reservoir Rocks

Material	Matrix density* (gm/cm^3)
Sand (consolidated)	2.65
Sand (unconsolidated)	2.65
Limestone	2.71
Dolomite	2.8–2.9
Shale	1.8–2.7
Gypsum	2.32
Anhydrite	2.9–3.0
Halite	2.16

From Reference 50.
*These figures are averages and may vary from area to area, depending on types and abundance of secondary minerals.

and sieving reservoir rock material. Fractions of the various sizes of grains are determined and described according to the nomenclature in Table 2.5. Larger grain size is associated with higher permeability, and very small grain sizes include sit and clay fractions that are associated with lower permeabilities.

2.2 DRILL STEM TESTS

A drill stem test (DST) is some form of temporary completion of a well that is designed to determine the productivity and fluid properties prior to completion of the well. Although a DST can be performed in uncased hole (open hole) or in cased hole (perforation tests), the open hole test is more common. The tool assembly which consists of a packer, a test valve, and an equalizing valve, is lowered on the drill pipe to a position opposite the formation to be tested. The packer expands against the hole to segregate the mud-filled annular section from the interval of interest, and the test valve allows formation fluids to enter the drill pipe during the test. The equalizing valve allows pressure equalization after the test so the packer can be retrieved. Details of the DST and DST assemblies are described elsewhere [13, 19] and will only be summarized here. By closing the test valve, a build up in pressure is obtained; by opening the test valve, a decline in pressure is obtained. During the DST, both pressures and flow rates are measured as a function of time.

Interpretation of DST results is often regarded as an art rather than a science. Certainly, a DST can provide a to valuable indication of commercial productivity from a well, provided engineering judgment and experience are properly utilized. Interpretations of various pressure charts are shown in Figure 2.1 [13, 17] and 2.2 [1]; details of interpreting DST data are described in the literature [18].

TABLE 2.4 Typical Core Analyses from Various Reservoirs [51] (a) Arkansas

Formation	Fluid prod.	Range of prod. depth, ft	Avg. prod. depth, ft	Range of prod. thickness, ft	Avg. prod. thickness, ft	Range of perm. K, md	Avg. perm. K, md	Range of porosity, %	Avg. porosity, %	Range of oil satn., %	Avg. oil satn., %	Range of calc. connate water satn., %	Avg. calc. connate water satn., %
Blossom	C/O*	2,190–2,655	2,422	3–28	15	1.6–8,900	1,685	15.3–40	32.4	1.2–36	20.1	24–55	32
Cotton Valley	C/O	5,530–8,020	6,774	4–79	20	0.6–4,820	333	11.3–34	20.3	0.9–37	13.1	21–43	35
Glen Rose†	O	2,470–3,835	3,052	5–15	10	1.6–5,550	732	17.3–38	23.4	4.0–52	21.0	28–50	38
Graves	C/O	2,400–2,725	2,564	2–26	11	1.2–4,645	1,380	9.8–40	34.9	0.3–29	16.8	19–34	30
Hogg	O	3,145–3,245	3,195	12–33	17	6.5–5,730	1,975	14.4–41	30.9	2.6–56	19.9	26–34	27
Meakin	G/C/O*	2,270–2,605	2,485	2–20	11	3.0–6,525	1,150	17.1–40	31.8	0.6–43	12.9	24–63	43
Nacatoch	C/O	1,610–2,392	2,000	6–45	20	0.7–6,930	142	9.9–41	30.5	0.2–52	4.9	41–70	54
Paluxy	O	2,850–4,890	3,868	6–17	12	5–13,700	1,213	15.1–32	26.9	7.5–49	21.2	28–43	35
Pettit	O	4,010–5,855	4,933	4–19	11	0.1–698	61	6.2–28	15.4	9.1–29	12.7	25–44	30
Rodessa†	O	5,990–6,120	6,050	8–52	16	0.1–980	135	5.1–28	16.5	0.7–26	14.8	25–38	31
Smackover§	G/C/O	6,340–9,330	8,260	2–74	18	0.1–12,600	850	1.1–34	14.2	0.7–41	12.8	21–50	31
Tokio	C/O	2,324–2,955	2,640	2–19	13	0.5–11,550	2,100	13.6–42	32.1	0.9–57	25.6	17–43	27
Travis Peak	C/O	2,695–5,185	3,275	3–25	10	0.4–6,040	460	9.4–36	24.3	0.5–36	14.3	16–48	36
Tuscaloosa	C/O	3,020–3,140	3,080	4–25	15	0.4–3,760	506	15.6–39	27.3	0.3–53	14.0	31–63	45

From Reference 51.

*Indicates fluid produced: G—gas; C—condensate; O—oil.

†Specific zone not indentified locally.

‡Includes data from Mitchell and Gloyd zones.

§Includes data from Smackover Lime and Reynolds zones.

TABLE 2.4 (b) East Texas Area

Formation	Fluid prod.	Range of prod. depth, ft	Avg. prod. depth, ft	Range of prod. thickness, ft	Avg. prod. thickness, ft	Range of perm. K, md	Avg. perm. K, md	Range of porosity, %	Avg. porosity, %	Range of oil satn., %	Avg. oil satn., %	Range of calc. connate water satn., %	Avg. calc. connate water satn., %
Bacon	C/O	6,685–7,961	7,138	3–24	11	0.1–2,040	113	1.5–24.3	15.2	2.7–20.6	8.6	9–22	16
Cotton Vally	C	8,448–8,647	8,458	7–59	33	0.1–352	39	6.9–17.7	11.7	1.1–11.6	2.5	13–32	25
Fredericksburg	O	2,330–2,374	2,356	5–8	7	0.1–4.6	1.2	11.9–32.8	23.1	3.3–39.0	20.8	35–43	41
Gloyd	C/O	4,812–6,971	5,897	3–35	19	0.1–560	21	8.0–24.0	14.9	tr–24.3	8.2	16–45	31
Henderson	G/C/O	5,976–6,082	6,020	3–52	12	0.1–490	19	7.0–26.2	15.2	0.8–23.3	10.6	21–44	27
Hill	C/O	4,799–7,668	5,928	3–16	9	0.1–467	70	6.4–32.2	15.6	0.9–26.7	12.2	23–47	33
Mitchell	O	5,941–6,095	6,010	3–43	21	0.1–487	33	7.2–29.0	15.5	1.8–25.9	12.5	15–47	29
Mooringsport	O	3,742–3,859	3,801	4–12	8	0.4–55	5	5.3–19.6	14.6	2.8–36.4	13.8	29–48	40
Nacatoch*	O	479–1,091	743	2–21	12	1.9–4,270	467	13.4–40.9	27.1	0.6–37.4	14.5	24–55	41
Paluxy	O	4,159–7,867	5,413	7–46	27	0.1–9,600	732	6.3–31.1	21.6	2.2–48.7	24.1	22–47	30
Pecan Gap	O	1,233–1,636	1,434	5–20	13	0.5–55	6	16.3–38.1	26.6	3.5–49.8	12.9	30–56	46
Pettit†	G/C/O	5,967–8,379	7,173	2–23	11	0.1–3,670	65	4.5–25.8	14.7	0.9–31.6	9.8	10–35	23
Rodessa	C/O	4,790–8,756	6,765	4–42	17	0.1–1,180	51	2.3–29.0	14.5	tr–25.3	5.3	6–42	23
Sub-Clarksville‡	O	3,940–5,844	4,892	3–25	12	0.1–9,460	599	8.2–38.0	24.8	1.4–34.6	17.9	12–60	33
Travis Peak§	C/O	5,909–8,292	6,551	2–30	11	0.1–180	42	5.6–25.8	15.0	0.1–42.8	12.5	17–38	28
Wolfe City	O	981–2,054	1,517	6–22	13	0.3–470	32	17.1–38.4	27.9	1.5–37.4	15.6	23–68	46
Woodbine	C/O	2,753–5,993	4,373	2–45	14	0.1–13,840	1,185	9.7–38.2	25.5	0.7–35.7	14.5	14–65	35
Young	C	5,446–7,075	6,261	4–33	17	0.1–610	112	4.4–29.8	19.7	tr–4.5	0.8	13–27	21

*Small amount of Navarro data combined with Nacatoch.
†Data for Pittsburg, Potter, and Upper Pettit combined with Pettit.
‡Small amount of Eagleford data combined with Sub-Clarksville.
§Data for Page combined with Travis Peak.

(Continued)

TABLE 2.4 (c) North Louisiana Area (*Continued*)

Formation	Fluid prod.	Range of prod. depth, ft	Avg. prod. depth, ft	Range of prod. thickness, ft	Avg. prod. thickness, ft	Range of perm. K, md	Avg. perm. K, md	Range of porosity, %	Avg. porosity, %	Range of oil satn., %	Avg. oil satn., %	Range of calc. connate water, %	Avg. calc. connate water satn., %
Annona Chalk	O	1,362–1,594	1,480	15–69	42	0.1–2.5	0.7	14.3–36.4	26.8	6.0–4.0	22	24–40	37
Buckrange	C/O	1,908–2,877	2,393	2–24	13	0.1–2,430	305	13.4–41	31.4	0.7–51	22.6	29–47	35
Cotton Valley[a]	G/C/O	3,650–9,450	7,450	4–37	20	0.1–7,350	135	3.5–34	13.1	0.0–14	3.1	11–40	24
Eagleford[b]	C	8,376–8,417	8,397	9–11	10	3.5–3,040	595	12.8–28	22.9	1.6–28	4.3	...	36
Fredericksburg	G/C	6,610–9,880	8,220	6–8	7	1.6–163	90	12.8–23.1	19.9	1.7–4.3	2.7	35–49	41
Haynesville	C	10,380–10,530	10,420	22–59	40	0.1–235	32	5.5–23.1	13.4	1.1–14.5	5.1	31–41	38
Hosston	C/O	5,420–7,565	6,480	5–15	12	0.4–1,500	140	8.8–29	18.6	0.0–35	8.8	18–37	28
Nacatoch	O	1,223–2,176	1,700	6–12	8	27–5,900	447	25.8–40	31.4	2.5–33	19.5	45–54	47
Paluxy	C/O	2,195–3,240	2,717	2–28	16	0.2–3,060	490	9.6–39	27.2	0.1–48	11.8	23–55	35
Pettit[c]	C/O	3,995–7,070	5,690	3–30	14	0.1–587	26	4.5–27	14.3	0.1–59	15.6	10–43	29
Pine Island[d]	O	4,960–5,060	5,010	5–13	9	0.2–1,100	285	8.5–27	20.6	13.3–37	24.1	16–30	22
Rodessa[e]	G/C/O	3,625–5,650	4,860	6–52	18	0.1–2,190	265	5.1–34	19.1	0.0–31	2.9	21–38	30
Schuler[f]	G/C/O	5,500–9,190	8,450	4–51	19	0.1–3,180	104	3.6–27.4	15.0	0.0–24	4.8	8–51	25
Sligo[g]	C/O	2,685–5,400	4,500	3–21	7	0.1–1,810	158	7.3–35	21.1	0.6–27	9.8	12–47	31
Smackover	C/O	9,980–10,790	10,360	6–55	24	0.1–6,190	220	3.4–23	12.9	1.1–22	7.2	9–47	25
Travis Peak[h]	C/O	5,890–7,900	6,895	7–35	18	0.1–2,920	357	7.0–27	19.4	0.1–35	8.6	26–38	31
Tuscaloosa	G/C/O	2,645–9,680	5,164	4–44	24	0.1–5,750	706	10.7–36	27.6	0.0–37	8.5	31–61	43

[a]Data reported where member formations of Cotton Valley group not readily identifiable.
[b]Data reported as Eutaw in some areas.
[c]Includes data reported as Pettit, Upper Pettit, and Mid Pettit. Sometimes considered same as Sligo.
[d]Sometimes referred to as Woodruff.
[e]Includes data reported locally for Jeter, Hill, Kilpatrick, and Fowler zones.
[f]Includes data reported locally Bodcaw, Vaughn, Doris, McFerrin, and Justiss zones.
[g]Includes data reported as Bridsong–Owens.
[h]Frequently considered same as Hosston.

TABLE 2.4 (d) California

Formation	Area	Fluid prod.	Range of prod. depth, ft	Ave. prod. depth, ft	Range of prod. thick-ness, ft	Ave. prod. thick-ness, ft	Range of perm. K, md	Ave. perm. K, md	Range of porosity, %	Ave. porosity, %	Range of oil satn., %	Avg. oil satn., %	Range of total water satn., %	Ave. total water satn., %	Range of calc. water satn., %	Ave. calc. connate water satn., %	Range of gravity, °API	Ave. gravity, °API
Eocene, Lower	San Joaquin Valley[a]	O	6,820–8,263	7,940	—	—	35–2,000	518	14–26	20.7	8–23	14.1	16–51	35	15–49	35	28–34	31
Miocene	Los Angeles Basin and Coastal[b]	O	2,870–9,530	5,300	60–450	165	10–4,000	300	15–40	28.5	6–65	18.8	25–77	50	15–72	36	15–32	26
Miocene, Upper	San Joaquin Valley[c]	O	1,940–7,340	4,210	10–1,200	245	4–7,500	1,000	17–40	28.2	9–72	32[k]	20–68[k]	50[k]	12–62	30	13–34	23
	Los Angeles Basin and Coastal[d]	O	2,520–6,860	4,100	5–1,040	130	86–5,000	1,110	19.5–39	30.8	10–55	25	22–72	44	12–61	30	11–33	21
Miocene, Lower	San Joaquin Valley[e]	O	2,770–7,590	5,300	30–154	76	15–4,000	700	20–38	28.4	4–40	19	25–80	51	14–67	36	15–40	34
	Los Angeles Basin and Coastal[f]	O	3,604–5,610	4,430	20–380	134	256–1,460	842	21–29	24.3	13–20	15.8	32–67	53	27–60	37	34–36	35
Oligocene	San Joaquin Valley[g]	O	4,589–4,717	4,639	—	—	10–2,000	528	19–34	26.3	12–40	22	2–60	43	3–45	30	37–38	38
	Coastal[h]	O	5,836–6,170	6,090	—	—	20–400	107	15–22	19.5	6–17	11.8	19–56	46	15–52	42	–	25
Pliocene	San Joaquin Valley[i]	O	2,456–3,372	2,730	5–80	33	279–9,400	1,250	30–38	34.8	7–43[l]	24.1[l]	33–84	54	10–61	34	18–44	24
	Los Angeles Basin and Coastal[j]	O	2,050–3,450	2,680	—	100	25–4,500	1,410	24–11	35.6	15–80	45	19–54	38	10–40	21	12–23	15

[a]Mainly data from Gatchell zone.
[b]Includes Upper and Lower Terminal, Union Pacific, Ford, "237," and Sesnon zones.
[c]Includes Kernco, Republic, and "26 R" zones.
[d]Includes Jones and Main zones.
[e]Includes "JV," Olcese, and Phacoides zones.
[f]Mainly data from Vaqueros zone.
[g]Mainly data from Oceanic zone.
[h]Mainly data from Sespe zone.
[i]Includes Sub Mulinia and Sub Scalez No. 1 and No. 2 zones.
[j]Includes Ranger and Tar zones.
[k]Oil-base data show high oil saturation (avg. 61 percent) and low water (3–54 percent, avg. 15 percent).
[l]Oil-base data show range 27.6 to 52.4 and avg. of 42.3 percent—not included in above "oil-saturation" values.

(Continued)

TABLE 2.4 (e) Texas Gulf Coast-Corpus Christi Area* (Continued)

Formation	Fluid prod.	Range of prod. depth, ft	Avg. prod. depth, ft	Range of prod. thickness, ft	Avg. prod. thickness, ft	Range of perm. K, md	Avg. perm. K, md	Range of porosity, %	Avg. porosity, %	Range of satn., %	Avg. oil satn., %	Range of calc. connate water satn., %	Avg. calc. connate water satn., %	Range of gravity, °API	Avg. gravity, °API
Catahoula	O	3,600–4,800	3,900	1–18	8	45–2,500	670	17–36	30	1–30	14	30–44	36	23–30	29
Frio	C/O	1,400–9,000	6,100	3–57	13	5–9,000	460	11–37	27	2–38	13	20–59	34	23–48	41
Jackson	O	600–5,000	3,100	2–23	9	5–2,900	350	16–38	27	3–32	15	21–70	45	22–48	37
Marginulina	C	6,500–7,300	7,000	5–10	7	7–300	75	14–30	24	1–4	2	20–48	34	55–68	60
Oakville	O	2,400–3,100	2,750	5–35	22	25–1,800	700	21–35	28	9–30	18	32–48	44	23–26	25
Vicksburg	C/O	3,000–9,000	6,200	4–38	12	4–2,900	220	14–32	24	1–17	7	26–54	38	37–65	48
Wilcox	C	6,000–8,000	7,200	30–120	60	1–380	50	15–25	19	0–10	1	22–65	37	53–63	58
Yegua	O	1,800–4,000	3,000	3–21	7	6–1,900	390	22–38	29	4–40	17	14–48	36	20–40	32

*Includes counties in Texas Railroad Commission District IV: Jim Wells, Brooks, Hidalgo, Aransas, San Patricio, Nueces, Willacy, Duval, Webb, Jim Hog, and Starr.

TABLE 2.4 (f) Texas Gulf Coast-Houston Area

Formation	Fluid prod.	Range of prod. depth, ft	Avg. prod. depth, ft	Range of prod. thickness, ft	Avg. prod. thickness, ft	Range of perm. K, md	Avg. perm. K, md	Range of porosity, %	Avg. porosity, %	Range of oil satn., %	Avg. oil satn., %	Range of total water satn., %	Avg. total water satn., %	Range of calc. water satn., %	Avg. calc. connate water satn., %	Range of gravity, °API	Avg. gravity, °API
Frio	C	4,000–11,500	8,400	2–50	12.3	18–9,200	810	18.3–38.4	28.6	0.1–6.0	1.0	34–72	54	20–63	34		
	O	4,600–11,200	7,800	2–34	10.4	33–9,900	1,100	21.8–37.1	29.8	4.6–41.2	13.5	24–79	52	12–61	33	25–42	36
Marginulina	C	7,100–8,300	7,800	4–28	17.5	308–3,870	2,340	35–37	35.9	0.2–0.8	0.5	33–61	46	14–31	21		
	O	4,700–6,000	5,400	4–10	5.7	355–1,210	490	28.5–37.3	32.6	8.1–21.8	15.3	48–68	59	25–47	36	25–30	28
Miocene	C	2,900–6,000	4,000	3–8	5.5	124–13,100	2,970	28.6–37.6	33.2	0.2–1.5	0.5	55–73	66	23–53	38		
	O	2,400–8,500	3,700	2–18	7.2	71–7,660	2,140	23.5–38.1	35.2	11–29	16.6	45–69	58	21–55	34	21–34	25
Vicksburg	C	7,400–8,500	8,100	1–6	2.0	50–105	86	26.5–31.0	27.1	0–1.5	0.2	66–78	74	53–61	56		
	O	6,900–8,200	7,400	3–18	9.3	190–1,510	626	29.5–31.8	30.4	14.4–20.3	15.3	45–55	53	26–36	35	22–37	35
Wilcox	C	5,800–11,500	9,100	5–94	19.1	3.0–1,880	96	14.5–27.4	19.6	0.2–10.0	1.5	27–62	46	20–54	38		
	O	2,300–10,200	7,900	3–29	10.0	9.0–2,460	195	16.2–34.0	21.9	4.6–20.5	9.7	32–72	47	20–50	37	19–42	34
Woodbine	O	4,100–4,400	4,300	6–13	8.2	14–680	368	23.5–28.7	25.5	10.7–27.4	20.1	34.4–72.7	46	24–59	36	26–28	27
Yegua	G/C	4,400–8,700	6,800	3–63	11.0	24–5,040	750	23.4–37.8	30.7	0.1–15.5	1.2	26–74	57	17–59	33		
	O	3,700–9,700	6,600	2–59	8.5	23–4,890	903	22.9–38.5	31.6	3.5–21.8	11.4	31–73	57	17–53	34	30–46	37
Louisiana Gulf Coast-Lafayette Area																	
Miocene	C	5,200–14,900	11,200	3–98	20.2	36–6,180	1,010	15.7–37.6	27.3	0.1–4.7	1.5	37–79	53	20–74	35		
	O	2,700–12,700	9,000	3–32	11.0	45–9,470	1,630	18.3–39.0	30.0	6.5–26.9	14.3	30–72	51	18–50	32	25–42	36
Oligocene	C	7,300–14,600	9,800	2–80	14.6	18–5,730	920	16.7–37.6	27.7	0.5–8.9	2.3	33–71	51	19–57	32		
	O	6,700–12,000	9,400	2–39	8.3	64–5,410	1,410	22.1–36.2	29.0	5.2–20.0	11.1	34–70	54	23–60	35	29–44	38

(Continued)

TABLE 2.4 (g) Oklahoma-Kansas Area (Continued)

Formation	Fluid prod.	Range of prod. depth, ft	Avg. prod. depth, ft	Range of prod. thickness, ft	Avg. prod. thickness, ft	Range of perm. K, md	Avg. perm. K, md	Range of perm. K_{90}, md	Avg. perm. K_{90}, md	Range of porosity, %	Avg. porosity, %	Range of oil satn., %	Avg. oil satn., %	Range of total water satn., %	Avg. total water satn., %	Range of calc. connate water satn., %	Avg. calc. connate water satn., %	Range of gravity, °API	Avg. gravity, °API
Arbuckle	G	2,700–5,900	4,500	5.0–37	18.3	3.2–544	131	—	—	9.0–20.9	14.4	0.7–9.4	3.7	34.5–62.7	43.1	28–62	40		
	O	500–6,900	3,500	1.0–65.5	11.8	0.2–1,530	140	0.1–1,270	67.8	2.1–24.3	12.0	5.2–42.3	17.7	20.6–79.3	52.4	20–79	47	29–44	37
	T[b]	800–11,600	3,600	2.0–33	14.3	0.1–354	57	0.1–135	21.8	3.7–23.1	9.2	0–23.8	7.1	37.2–91.9	69.2	37–91	52	42	42
Atoka[c]	G	3,700–3,800	3,700	1.0–9.0	4.0	1.3–609	174	—	—	8.5–17.3	12.9	0–8.1	2.0	36.4–65.2	47.2	32–65	45		
	O	500–4,500	2,600	3.0–16	7.8	0.3–920	144	0.6–2.8	1.7	5.9–28.6	14.5	5.1–35.1	20.7	16.4–61.5	38.7	19–61	37	31–42	38
	T	300–3,700	2,100	2.0–10	6.5	9–166	67.3	—	—	11.9–18.6	14.9	5.8–21.1	12.1	42.7–55.4	47.0	40	40		
Bartlesville	G	700–7,400	2,600	1.5–42	11.4	0.2–36	10.4	5.5	5.5	8.4–21.1	15.6	0–11.1	4.7	23.4–70	54.1	23–68	48		
	O	200–5,700	1,500	1.0–72	14.0	0.2–537	32.7	1.5	1.5	8.5–25.8	17.8	3.3–60.6	18.2	17.4–85.2	44.4	17–72	40	28–42	34
	T	500–2,600	1,200	4–40	14.5	0.1–83	18.2	0.07	0.07	8.5–20.1	14.6	0.9–35.7	12.2	43.9–88	63.5	43–67	54	35	35
Bois D'Arc	G	4,800–5,100	5,000	4–48	19	0.1–43	24.4	0.1–2.2	0.45	3.8–19.8	12.2	0–8.7	4.3	32.9–62.4	42.8	26–62	40	32–42	40
Booch	G	3,700–7,800	6,500	2.3–50	12.5	0.3–664	36.0	—	—	1.2–19.3	7.2	3.3–25.8	15.0	14.6–58.5	32.4	15–59	32		
	O	2,600–3,200	2,900	5–8	6.5	1.4–6.6	4.0	—	—	11.9–14.8	13.4	4.6–8.8	6.7	50–51.3	50.7	50	50		
	T	1,000–3,800	2,600	2–26.5	8.8	0.3–160	19.3	—	—	8.3–21.4	15.6	4.8–49.7	21.5	15.3–60	40.0	15–59	37	29–42	35
	T	2,700–3,300	3,000	4–5	4.5	3.1–13	8.0	—	—	16.9–18.1	17.5	7.4–7.8	7.6	47.3–55.2	51.3	44	44		
Burgess	G	—	1,600	—	20	—	142	22	22	—	14.2	—	6.3	—	37.3	—	35		
	O	300–2,800	1,800	2.5–9	5.8	0.2–104	19	0.4	0.40	8.1–22.8	13.2	16.2–33	21.5	19.3–65.4	42.2	19–58	40	31–38	36
First Bromide[d]	G	6,800–7,600	7,200	3.0–19.5	11.3	0.6–62	31.3	—	—	1.5–6.5	4.0	0–7.6	3.8	35.7–71.8	53.8	36–72	54		
	O	3,700–13,800	8,600	2.0–82	18.7	0.1–2,280	175	0.2–7.4	2.23	1.4–15.7	9.8	3.1–24	11	12.8–67.2	35.4	12–67	34	31–42	40
	T	6,000–13,200	11,500	15–161.3	65.1	0.9–40	18.3	1.40	1.40	1.5–10.9	6.5	0.4–6.8	2.2	29.5–78.8	48.3	28–45	32		
Second Bromide[e]	G	6,900–16,200	12,800	20–53.6	37.9	3.4–72	21.4	0.3–0.9	0.60	3.5–14.5	6.8	0–6.9	4.0	28.2–45.7	37.9	8–44	25	42	42
	O	4,500–11,200	9,000	3.0–6.9	16.2	2.0–585	118	—	—	5.6–11.7	9.3	2.4–24.2	11.5	8.9–44.9	25.1			37–42	41
	T	4,400–13,300	9,700	5–44.5	18.4	0.8–42	12.9	—	—	5.8–11.4	7.4	0–13.6	4.8	21.1–57.6	43.5	40			
Burbank	O	1,300–4,500	2,800	3–48	17.3	0.1–226	8.64	—	—	8.4–21.6	15.7	9.3–26.6	15.3	31.5–73.4	47.2	31–73	43	35–41	39
	T	2,800–3,700	3,000	3–19	9.1	0.1–4.8	1.53	—	—	7.1–17.0	13.7	2.0–15.7	11.2	45.7–80.7	57.8	45–81	51		

Chester	G	4,200–6,700	5,700	2–45	10.9	0.1–269	33.0	0.9–3.5	1.87	2.6–20.7	12.2	0–7.5	1.1	20.9–80.7	46.8	19–81	43		
	O	4,700–6,700	5,700	2–23	8.6	0.1–61	9.11	0–0.5	0.21	2.3–16.0	10.1	7.2–35.9	19.1	17.7–80.8	42.1	17–81	33	38–42	40
	T	4,800–6,100	5,700	4–20.5	10.0	0.1–13	2.38	0.1–5.0	1.18	3.2–17.8	7.7	0–11.1	1.2	40–89.2	61.7	40–89	61		
Clevland[f]	G	2,200–5,700	3,500	2–17	9.0	2.5–338	50.6	–	–	9.8–23.5	16.9	0–7.1	4.1	40–64.4	48.9	30–64	42		
	O	300–6,400	3,200	1–70	13.4	0.1–135	15.4	1.4–2.3	1.85	7.4–24.6	15.2	5.8–35.5	13.1	10.2–74.0	46.7	10–74	44	27–56	42
	T	1,900–3,900	3,100	3–22	7.7	0.1–112	12.9	–	–	11.0–20.4	15.6	0–21.1	7.8	39.9–77.2	55.3	32–77	49		
Deese[g]	G	4,300–11,800	6,500	5–55	19.3	7.8–232	94.1	0.80	0.80	9.8–22.6	16.7	2.2–6.3	3.8	19.1–54.9	42.1	19–49	37		
	O	600–10,000	5,200	2–60.3	11.7	0.4–694	62.8	1.10	1.10	4.7–26.4	17.4	5.9–46.4	20.4	14.0–56.6	37.8	13–57	33	17–42	32
Hoover	G	2,200–6,800	4,000	4–49	16.6	1.9–200	61.8	–	–	11.7–23.4	18.3	0–7.0	0.8	41.1–77.1	53.8	19–76	45		
	O	1,800–2,100	2,000	3–37	11.9	1.3–974	288	–	–	12.7–24.1	19.7	12.6–23.1	16.0	14.6–48.5	40.2	14–47	35	36–42	42
	T	1,900–2,000	2,000	2–17	8.4	55–766	372	–	–	16.7–22.5	20.5	6.6–17.1	14.5	34.8–50.7	42.9	31–42	35		
Hoxbar	G	3,800–8,880	6,300	9–11	10.0	6.4–61	33.7	–	–	13.9–18.2	16.1	0.7–4.4	2.6	40.1–40.6	40.4	34–39	37	42	42
	O	1,000–10,300	4,200	2–63	14.4	0.1–1,620	277	–	–	3.1–29.7	18.5	3.2–48.7	21.4	13.8–68.5	45.1	13–68	39	29–42	34
	T	2,900–3,000	3,000	3–13	9.3	0.5–31	14.4	–	–	14.3–22.7	18.5	3.3–11.4	6.6	50.5–69.8	57.9				
Hunton	O	1,800–9,600	4,600	2–77.3	14.0	0.1–678	34.5	0–77.0	5.24	1–33.8	10.9	1.6–34.5	15.3	16.7–93.4	48.6	17–93	46	24–42	38
	T	2,500–8,700	4,900	2–73	14.7	0.1–48	5.3	0.1–7.9	2.04	1–19.5	7.3	0–61.1	10.6	16.0–88.7	54.5	16–89	48		
Lansing	O	1,900–5,800	3,800	3–16.2	6.5	0.3–390	101	0.3–162	52.3	8.4–16.0	12.2	6.5–28.9	18.1	37.4–68.6	51.9	28–69	49	31–39	37
	T	–	3,300	–	22.0	–	14	–	6.7	–	7.2	–	12.8	–	75.5				
Layton	G	700–6,100	3,900	4–18	9.3	0.2–210	26.3	–	–	5.1–25.9	14.5	0–7.8	2.4	38.2–83.7	54.1	34–83	47		
	O	500–6,300	2,900	1–57	10.3	0.3–280	54.1	0.5–162	23.3	4.6–27.2	17.8	1.6–37.3	15.3	28–76.3	45.5	23–76	41	30–42	37
	T	1,800–5,700	3,200	1.5–15.5	7.4	1.1–143	23.8	–	–	14.2–21.3	17.1	0–14.3	6.9	33.2–69.4	45.9	31–69	43		
Marmatom	O	4,300–4,600	4,400	1.5–7.5	4.7	24–105	46.4	0.20	0.20	1.8–21.4	14.0	6.4–16.1	11.7	42.8–66.4	55.5	42–66	53	36–42	40
Misner	G	8,100	8,100	3–14	8.5	37–171	104	–	–	11.0–12.1	11.6	2.1–2.3	2.2	19.8–22.9	21.4	18–22	20		
	O	2,600–6,500	4,300	2–56.5	10.6	0.1–803	89.7	0–2.1	0.62	2.1–20.9	11.9	4.1–41.6	14.8	16.9–86.7	41.5	14–87	38	36–48	42
	T	4,900–6,200	6,000	8–21	15.8	0.1–120	41.8	–	–	1.9–11.3	8.1	0–8.2	4.7	21.4–51.7	33.0	20–51	32		

(Continued)

TABLE 2.4 (g) Oklahoma-Kansas Area (Continued)

Formation	Fluid prod.	Range of prod. depth, ft	Avg. prod. depth, ft	Range of prod. thickness, ft	Avg. prod. thickness, ft	Range of perm. K, md	Avg. perm. K, md	Range of perm. K_{90}, md	Avg. perm. K_{90}, md	Range of porosity %	Avg. porosity %	Range of oil satn, %	Avg. oil satn, %	Range of total water satn, %	Avg. total water satn, %	Range of calc. connate water satn, %	Avg. calc. connate water satn, %	Range of gravity, °API	Avg. gravity, °API
Mississippi Chat	G	1,800–5,100	4,000	2–34.4	16.1	0.4–516	33.5	0.2–74	13.9	6.5–37.8	21.0	0–6.8	2.4	60.3–93.4	76.7	60–93	77		
	O	800–5,200	3,100	2–48.1	12.2	0.1–361	21.9	0–216	13.7	5.7–39.3	22.3	1.4–30.0	12.9	27.1–94.8	64.0	27–95	58	22–42	35
	T	1,200–5,200	3,900	1–43	10.9	0.2–229	21.3	0–163	14.2	1.5–38.0	18.7	1.1–18.3	7.6	47.4–84.9	71.5	43–85	63		
Mississippi Lime	G	900–8,800	4,600	3–27.1	13.3	0.1–129	22.2	0.1–89	13.2	1.5–23.6	10.3	0–9.3	2.8	22.6–93.5	63.2	22–93	53		
	O	600–6,600	4,100	1.5–95.3	12.0	0.1–1,210	43.5	0.1–185	9.44	1.3–34.1	13.4	2.1–56.5	15.0	16.9–85.3	50.7	16–85	46	22–45	39
	T	400–7,200	4,000	4–70.1	17.4	0.1–135	7.5	0.1–36	4.23	1.1–26.1	9.3	0–41.2	6.9	32.9–94.0	67.6	32–94	61		
McLish	G	3,600–17,000	10,100	14–58	35.3	12–98	48.0	—	—	2.8–9.6	6.7	4.0–14.7	7.8	19.3–76.5	43.9	19–77	44		
	O	1,600–11,200	8,100	3–42	12.2	0.7–157	39.0	—	—	5.5–1.5	11.0	5.1–27.7	13.2	14.8–52.2	32.1	14–52	31	35–48	38
Morrow	G	4,300–9,700	6,100	2–64	11.0	0.1–1,450	115	6.2–8.8	7.5	4.2–24.4	14.8	0–33.0	4.3	29.0–77.0	46.5	16–77	36		
	O	4,100–7,500	5,700	2–37	9.8	0.2–1,840	117	0.3–55	23.1	57–23.2	14.6	0.7–44.5	15.1	23.9–75.5	42.1	16–54	35	33–43	40
	T	5,500–6,900	6,100	3–30	9.5	0.1–410	34.4	0.1–48	28.0	5.5–16.2	11.3	0–15.2	5.0	31.1–90.1	57.2	31–90	38		
Oil Creek	G	7,100–14,000	10,900	14–149	46.3	0.1–132	32.0	—	—	6.1–13.5	9.0	0–6.5	1.6	12.5–40.6	25.2	12–40	24		
	O	5,100–11,700	8,300	3–71	12.6	0.1–615	131	0.2–230	75.6	1.8–23.9	13.1	1.3–29.5	13.0	14.2–76.4	39.1	14–76	34	29–42	36
	T	8,400–13,700	12,300	8–27	15.0	0.1–87	22.1	—	—	5.2–16.1	10.9	0–5.8	2.6	21.7–74.9	46.6	21–74	46		
Oswego	G	4,500–4,600	4,600	8–9	8.5	2.4–151	76.7	—	—	12.0–17.3	14.7	5.1–6.4	5.8	39.8–55.5	47.7	34–55	45		
	O	300–6,300	3,800	3.6–34.1	12.3	0.2–296	27.3	0.1–66	9.24	2.6–21.6	10.1	0–27.1	15.0	16.2–73.4	41.5	15–73	37	35–48	44
	T	1,200–5,800	3,300	2–21	10.6	0.1–117	27.0	0–41	11.5	4.7–20.9	8.7	0–14.5	5.8	41.7–89.7	63.4	42–89	57		
Peru	G	1,200–5,300	3,100	4–17	9.8	3.1–42	15.0	—	—	12.3–17.5	15.6	0.1–7.9	4.1	44.3–59.4	52.5	44–56	51		
	O	200–3,200	1,200	2–42	12.4	0.2–284	20.8	—	—	12.7–33.8	18.7	6.7–36.8	14.7	34.4–73.1	50.6	28–73	44	25–43	36
	T	700–2,500	1,500	4–21	10.3	1.7–804	205	—	—	13.6–24.4	19.2	2.8–25.5	12.0	38.0–60.4	50.7	36–56	51		
Prue	G	3,000–6,600	4,000	5–22	13.8	0.7–42	18.3	—	—	13.8–22.4	17.8	2.3–9.1	5.5	31.4–53.4	42.2	25–49	37		
	O	600–6,700	3,100	2–81	14.6	0.1–254	22.6	—	—	7.6–23.8	17.0	4.7–34.1	16.9	24.4–73.1	41.6	20–72	38	34–46	42
	T	3,000–5,400	3,700	3–18	11.7	0.5–133	42.8	—	—	9.8–23.4	17.5	3.7–34.3	19.0	40.7–60.9	47.1	32–60	36		
Purdy	O	4,200–7,400	4,500	3–30	14.8	7.4–500	182	51–266	179	12.3–18.8	16.7	10.1–27.2	20.0	31.4–58.1	41.5	16–50	29	39–44	41
	T	—	4,200	—	4.8	—	195	—	166	—	17.8	—	13.6	—	56.2	—			

Reagan	G	3,500–3,600	3,600	2–13	7.4	1.1–173	39.3	—	—	9.3–12.7	10.8	1.1–7.9	4.2	28.4–68.4	44.2	28–68	40	41	41
	O	2,100–3,700	3,600	1–32	11.0	0.2–2,740	255	—	—	6.9–21.5	13.3	3.0–42.0	14.2	17.5–72.9	32.9	12–72	31	24–43	38
Redfork	T	3,600	3,600	5–7	6.0	19.0–37	38.0	—	—	10.6–12.8	11.7	1.8–10.5	6.2	33.3–46.7	40.0	29–45	29		
	G	2,300–7,400	4,300	4–19	7.9	0.1–160	23.4	—	—	3.8–21.2	14.5	0–21.7	4.7	16.2–63.6	45.8	16–63	39	32–48	37
	O	300–7,600	3,100	1–63	10.5	0.1–668	14.2	—	—	6.6–26.1	16.2	5.4–30.8	16.9	29.5–57.7	43.7	27–55	41		
Skinner	G	1,200–3,800	3,100	2–9	5.3	0–23	6.3	—	—	10.1–18.6	15.3	0.3–36.3	9.9	41.4–69.7	52.6	41–69	49	30–46	36
	O	1,000–5,300	3,700	4–29	11.8	0.1–127	27.7	—	—	13.3–19.8	15.7	0–9.9	4.2	30.6–48	40.8	26–47	38		
	O	1,000–5,800	3,200	1–42.5	9.2	0.1–255	20.6	2–6.6	3.30	7.4–21.7	15.3	2.5–29.7	20.1	14.3–78.7	40.3	14–78	38		
	T	2,400–4,600	3,400	6–35.9	11.5	0.3–16	6.0	2.40	2.40	11.7–19.0	15.5	4.9–18.2	8.5	39.9–71.1	52.4	39–71	39		
Strawn	G	—	1,100	—	12.0	—	71.0	—	—	—	21.3	—	9.9	—	61.8	—	38		
Sycamore	G	1,000–7,400	3,500	2–40.5	12.4	0.1–599	58.1	—	—	8.2–23.5	16.8	5.7–31.1	15.1	28.5–61.5	45.6	22–56	41	31–44	40
	O	2,600–6,700	4,600	2–84	26.4	0.1–3.1	0.67	0.13	0.50	7.2–21.4	13.3	9.2–33.5	21.1	36.0–61.6	45.5	32–62	43	33–36	35
Tonkawa	G	5,000–7,100	5,600	4.42–27.5	9.8	0.3–283	46.7	—	—	11.7–21.4	16.4	0–8.1	2.0	31.6–56.3	44.5	27–56	41	40–45	43
	O	2,400–5,700	4,800	4.42–28.5	8.7	1.4–278	98.6	8–22	15.0	13.2–22.9	18.4	7.5–16.5	12.5	36.1–78.0	45.0	31–78	38		
Tucker	O	2,300–3,100	2,700	4–9	7.0	1.3–406	106	—	—	15.4–18.9	17.1	6.9–17.3	11.4	45.1–52.6	49.0	44–52	45		
Tucker	O	1,300–2,900	2,200	2–14	7.6	2.1–123	36	—	—	12.4–20.3	15.6	7.3–29.8	16.0	35.6–50.1	40.7	33–43	38	29–40	36
	T	2,700–2,900	2,800	8.9–16	12.5	4.3–252	128	53	53	11.8–19.5	15.7	7.1–10.9	9.0	58–64.3	61.2	52–62	52		
Tulip Creek	G	7,200–16,700	13,400	21–268.4	78.1	0.9–24	7.63	0.5–1.0	0.40	2.0–11.9	6.1	0–6.6	4.1	23.7–54.8	33.2	23–55	34	49.5	49.5
	O	700–16,800	8,000	2–136	15.3	0.1–1,470	154	0.2–1.8	0.80	2.5–25.0	11.6	3.0–44.5	12.2	10.0–63.0	34.9	9–63	33	32–50	40
Viola	G	1,400–12,900	8,600	3–86.5	20	2.0–143	44.6	0.40	0.40	0.7–26.0	11.0	0.7–7.7	2.6	15.9–82.8	45.7	15–82	46		
	G	4,300–7,300	5,400	3–73	39.1	3.6–23	10.8	3.40	3.40	8.1–10.1	9.3	1.7–9.4	5.0	19.7–37.2	30.7	19–37	30		
	O	2,100–11,100	4,900	2–111.7	17.2	0.1–1,150	52.3	0.2–186	18.3	1.0–16.1	8.4	3.2–41.0	15.5	24.1–85.5	54.4	24–86	51	28–48	37
	T	2,600–10,300	4,600	2–117	19.6	0.1–997	45.1	0.3–49	4.38	0.6–18.8	7.1	0–33.7	8.6	39.0–90.8	65.7	39–90	58		
Wayside	O	300–2,800	800	3.1–34	10.8	0.2–133	22.2	—	—	13.2–24.9	18.6	8.1–33.8	18.6	29.4–68.0	51.3	28–67	47	29–42	35

(Continued)

TABLE 2.4 (g) Oklahoma-Kansas Area (Continued)

Formation	Fluid prod.	Range of prod. depth, ft	Avg. prod. depth, ft	Range of prod. thickness, ft	Avg. prod. thickness, ft	Range of perm. K, md	Avg. perm. K, md	Range of perm. K_{90}, md	Avg. perm. K_{90}, md	Range of porosity, %	Avg. porosity, %	Range of oil satn, %	Avg. oil satn, %	Range of total water satn, %	Avg. total water satn, %	Range of calc. connate water satn, %	Avg. calc. connate water satn, %	Range of gravity, °API	Avg. gravity, °API
First Wilcox	G	2,800–5,400	4,300	2–35	11.3	0.7–145	72.1	—	—	5.2–15.6	10.8	0.7–8.3	3.6	29.7–60.5	43.9	29–60	44		
	O	2,800–7,400	4,900	2–28	10.0	0.2–445	91.3	—	—	5.4–20.5	12.0	3.6–40.5	11.7	15.0–58.2	32.0	14–58	31	33–50	42
	T	3,200–6,100	3,900	1.9–29	7.7	0.3–418	84.1	0.80	0.80	6.8–17.7	10.9	0–16.9	7.9	24.8–63.6	41.7				
Second Wilcox	G	5,000–10,000	6,700	5–28	13.4	0.2–154	76.2	—	—	5.0–15.1	11.2	0–3.8	1.5	17.7–45.8	30.9	17–43	29		
	O	3,700–8,400	6,500	1.3–32	11.3	0.4–2,960	214	—	—	4.2–20.6	12.4	2.9–19.2	10.2	19.0–56.3	36.9	18–56	34	34–42	40
	T	4,700–7,500	6,000	1.5–5	4.4	0.4–756	246	—	—	1.9–20.4	12.9	0–8.4	6.1	41.4–60.5	42.5	40–60	43		
Woodford	O	4,100–5,000	4,600	2.6–30.4	16.2	1.4–250	87.1	2.4–156	79.2	1.9–6.6	4.4	8.3–16.7	11.8	43.0–87.9	60.1	43–87	60	41	41

[a]General geologic sections taken at different points in Oklahoma-Kansas areas indicate some variations in the properties and an appreciable variations in the occurrence and relative depths of many of the more important oil- and/or gas-producing zones, formations, geologic groups, and their members. The general identification of core samples from these producing intervals reflect local conditions or activities significantly. In the development of the average data values, an attempt has been made to combine data originally reported for locally named zones into more generally recognized formations or geologic groups. In some instances (i.e., Deese, Cherokee) data are reported for a major geologic group as well as for some of its individual members. The values designated by the major group name represent areas where the general characteristics permit identification as to the geologic group but not as to group members. In other areas the group members or zones are readily identifiable. The combinations of data and the use of local rather than regional geologic names in some instances are explained in the footnotes. [b]T Represents transition zone or production of both water and either gas or oil.
[c]Includes data reported as Dornick Hills and Dutcher.
[d]Includes Bromide First and Second as reported on McClain Country area.
[e]Data reported locally as Bromide Third, Bromide Upper third, and Bromide Lower have been considered as part of the Tulip Creek.
[f]Includes data reported as Cleveland Sand, Cleveland, Lower, and Cleveland Upper.
[g]Includes the numerous zones (Deese First, Second, Third, Fourth, Fifth, Zone A, Zone B, Zone C, and Zone D) reported locally for the Anadarko, Ardmore, and Marietta Basin areas. In northwest Oklahoma, these different zones are normally refereed to as Cherokee. In other areas the zones are frequently identifiable and properties are reported as for Redfork, Bartlesville, etc.

TABLE 2.4 (h) Rocky Mountain Area

Formation	Fluid prod.	Range of prod. depth, ft	Avg. prod. depth, ft	Range of prod. thickness, ft	Avg. prod. thickness, ft	Range of perm. K, md	Avg. perm. K, md	Range of perm. K90, md	Avg. perm. K90, md	Range of porosity, %	Avg. porosity, %	Range of oil satn. %	Avg. oil satn. %	Range of total water satn. %	Avg. total water satn. %	Range of calc. connate water satn. %	Avg. calc. connate water satn. %	Range of gravity, °API	Avg. gravity, °API
Aneth	O	5,100–5,300	5,200	3.8–23.1	14.0	0.7–34	9.35	0.2–23	6.10	4.4–10.5	8.1	14.5–35.9	25.0	12.5–30.5	23.6	13–31	24	41	41
Boundary	G	5,500–5,600	5,600	8–27	17.5	0.1–2.0	1.05	—	—	4.3–6.5	4.7	4.7	4.7	23.8–35.0	29.4	23–35	29		
butte	O	5,400–5,900	5,600	2–68	16.2	0.1–114	13.3	0.2–33	12.5	5.4–21.6	11.0	4.8–26.7	12.5	9.3–48.8	28.3	7–45	27	40–41	41.1
Cliffhouse	G	3,600–5,800	4,800	2–58	13.7	0.1–3.7	0.94	—	—	7.0–16.2	11.3	0–19.8	4.5	10.2–60.3	36.9	10–59	36		
D Sand	O	4,350–5,050	5,800	7–33	15.0	0–900	192	—	—	8.6–29.5	21.6	8.4–39.5	13.2	14.8–55.3	40.6	9–48	23	36–42	38
Dakota	O	500–7,100	5,700	2–24	9.5	0.1–710	28.6	0.08	0.08	7.3–19.6	11.2	0–7.8	3.5	11.6–44.3	31.0	11–44	29	38–43	40
Desert	O	3,400–7,200	5,600	4–19	7.9	0.1–186	22.3	33.0	33	5.0–23.3	12.7	13.8–35.9	24.4	14.8–24.7	19.2	14–25	19	41	41
	O	5,400–5,500	5,500	11.6–18.3	14.9	1.0–11	4.4	0.4–2.4	1.13	11.9–13.8	12.7	13.4–16.8	15.2	—	—	28–45	33	31–50	41
Frontier sands	O	265–8,295	2,950	8–100	46	0–534	105	—	—	6.3–29.8	20.0	7.6–37.6	14.9	20.7–59.2	40.0	20–54	37	39	39
Gallop	O	1,500–6,900	5,000	5–25	11.6	0.1–324	26.5	0.3–20	10.2	8.5–20.8	13.3	0–25.6	5.7	17.2–76.9	35.7	14–77	34	36–42	39
Hermosa	O	500–6,400	4,600	2–43	12.4	0.1–2,470	48.2	0.1–3.2	0.7	6.9–23.1	12.5	8.5–43.7	25.3	14.2–45.3	32.7	12–45	32	41–42	40
Hospa	O	4,900–7,700	5,600	5–30	14.1	0.1–91	18.6	45.0	45.0	5.5–16.5	10.2	0–6.5	3.0	11.6–60.0	35.6	12–60	35	40	40
J Sand	O	5,300–6,000	5,600	3–38.2	15.1	0.1–37	7.32	0–26	4.26	2.7–17.9	8.3	3.9–29.1	10.8	8.7–49.7	38.1	8–49	37	36–42	38
	O	4,800–7,100	5,500	3–17	10.5	0.1–70	18.2	—	—	7.4–11.9	10.5	0.5–23.8	7.5	32.3–44.8	36.0	31–45	35		
	O	4,600–5,100	4,800	6–18	13.3	0.7–25	8.63	—	—	6.6–14.8	11.3	20.4–29.8	25.0	—	—	6–42	20		
	O	4,470–5,460	4,900	15–62	25	0–1,795	330	—	—	8.9–32.7	19.6	8.8–46.5	13.9	—	—	22–33	27		
Madison lime†	O	3,400–6,200	4,900	41–450	186	0.1–1,460	13	—	—	1.8–26.4	11.9	6.0–43.5	17.4	—	—	15–43	27		
Menefee	G	5,200–5,700	5,400	7–25	12.7	0.1–20	5.03	—	—	8.7–13.5	11.2	0.3–5.3	1.6	14.5–45.1	27.5	15–64	40		
Mesa Verde	G	1,500–6,100	4,700	2–22	10.0	0.1–17	3.57	—	—	10.0–19.8	14.6	0–6.8	3.3	14.5–68.4	42.0	—	44		
	O	*	300	—	4.0	—	60	—	—	—	26.2	—	8.3	—	61.0	15–41	35	21.6–30	26
Morrison	O	1,600–6,900	4,500	24–54	40	0–1,250	43	—	—	9.9–25.5	17.5	5.0–26.0	13.1	—	—	5–47	19	29–56	42
Muddy	O	930–8,747	1,845	7–75	20	0–2,150	173	—	—	2.3–32.9	22.3	7.6–48.5	30.8	—	—	10–58	34	26–42	38
Paradox	G	5,100–9,900	6,900	4–44.2	12.2	0.2–42	11.6	0.1–28	4.43	1.4–19.4	7.4	0–10.1	3.1	9.9–57.9	34.7	10–61	33		
	O	5,300–6,100	5,700	2–66	14.8	0.1–119	10.4	0–57	4.57	3.3–21.8	10.5	3.6–38.7	12.4	10.8–60.5	33.6			40–43	41

(Continued)

TABLE 2.4 (h) Rocky Mountain Area (Continued)

Formation	Fluid prod.	Range of prod. depth, ft	Avg. prod. depth, ft	Range of prod. thickness, ft	Avg. prod. thickness, ft	Range of perm. K, md	Avg. perm. K, md	Range of perm. K90, md	Avg. perm. K90, md	Range of porosity, %	Avg. porosity, %	Range of oil satn., %	Avg. oil satn., %	Range of total water satn., %	Avg. total water satn., %	Range of calc. connate water satn., %	Avg. calc. connate water satn., %	Range of gravity, °API	Avg. gravity, °API
Phoshporia (formerly Embar)	O	700–10,500	4,600	5–100	64	0–126	3.7	—	—	2.0–25.0	8.9	3.0–40.0	22.5	—	—	5–30	21	15–42.3	25.4
Picture Cliffs	G	1,800–5,800	3,400	3–43	16.8	0.1–7.68	1.12	—	—	8.5–25.5	15.7	0–21.1	2.6	23.6–67.5	46.7	23–53	43	55	55
	O	*	2,900	—	23.0	—	0.5	—	—	—	11.4	—	23.3	—	49.1	—	49		
Point Lookout	G	4,300–6,500	5,500	2–101	22.9	0.1–16	1.74	—	—	5.6–21.6	10.9	0–9.1	2.9	11.9–55.6	36.7	12–55	36	—	39
			4,700	—	7.0	—	2.90	—	2.40	—	13.3	—	23.8	40	40.9	—	41		
Sundance	O	1,100–6,860	3,100	5–100	44	0–1,250	100	—	—	15.0–25.0	19.0	8.0–25.0	17.0	7.0–59	—	20–49	35	22–63	39
Tensleep	O	600–11,800	4,700	10–200	118	0–2,950	120	—	—	5.0–27.0	13.6	6.0–30.0	23.3	—	25.7	5–50	19	17–58.5	26.2
	G	*	7,900	—	7	—	230	—	—	—	20.2	—	4.0	—	51.8	—	43		
Tocito	O	1,400–5,100	4,600	4–58	17.3	0–31	3.36	—	—	12.8–17.8	14.7	11.9–26.6	21.3	40.8–55	46.3	40–55	46	36–40	36

*Not enough wells to justify range of variations.

†Data limited to Big Horn Basin.

TABLE 2.4 (i) West Texas-Southeastern New Mexico Areas

Formation	Group[a]	Range Area[b]	Fluid prod. depth prod.	Range of prod. depth, ft	Avg. prod. depth, ft	Range of prod. thick-ness, ft	Avg. prod. thick-ness, ft	Range of perm. K, md	Avg. perm. K, md	Range of perm. K90, md	Avg. perm. K90, md	Range of porosity, %	Avg. porosity, %	Range of oil satn., %	Ave. oil satn., %	Range of total water satn., %	Avg. total water satn., %	Range of connate water satn., %	Avg. connate water satn., %	Range of gravity, °API	Avg. gravity, °API
Bend conglomerate	1	1, 2, 4, 5, 6, 8	O	6,000–6,100	6,000	3–22	13.2	4–311	150	—	—	13.8–16.9	15.0	8.1–8.6	8.3	43–64	52	42–62	50	40–42	41
	2	3	O	10,300–10,500	10,400	10–28	20.0	1.6–11	5.7	0.9–5.1	2.2	4.0–15.7	10.9	9.5–16.1	11.5	21–41	33	21–39	32	41–45	43
Blinebry	—	8	G	5,383–5,575	5,480	23–50	36	1.6–3.8	2.4	—	—	10.7–14.8	12.7	2.1–9.1	4.9	34–40	36	31–33	33	31–33	33
Cambrian	—	3	O	5,262–5,950	5,610	4–95	43	0.1–5.3	1.8	0.2–4.2	1.4	3.1–12.5	7.8	9.6–19.3	12.8	29–57	40	27–56	39	39–42	40
Canyon Reef	—	2, 3, 4, 5, 6, 7	O	5,500–6,300	5,900	2.0–95	30.3	0.8–1,130	173	0.3–249	17	4.1–16.8	12.0	6.9–21.2	11.8	22–71	39	22–71	38	44–51	48
Canyon Sand	—	2, 3, 4, 7	G	4,200–10,400	7,100	4.0–222	36.8	0.6–746	42	—	—	3.0–21.5	8.9	3.6–39.2	11.6	18.3–73	44	18–73	43	30–47	42
	—		O	—	5,000	—	8.0	—	1.7	—	—	—	15.1	—	5.8	—	46	—	44	—	—
Clearfork	1	3, 4, 7	G	3,000–10,000	5,500	3.0–57	16.9	0.1–477	38	—	7.8	5.5–22.1	14.3	4.8–27.7	13.7	21–72	43	21–72	41	37–43	40
	—		O	—	2,400	—	95	—	11	—	—	—	13.5	—	5.7	—	50	—	50	—	—
	2 (part)	5, 6	O	1,500–6,800	4,400	4.0–180	41	0.1–43	4.6	0.1–24	2.5	4.1–20.6	9.2	7.5–31.4	15.6	18–84	54	18–84	53	23–42	28
	2 (part)		O	5,400–8,300	6,600	3.0–259	3.3	<0.1–136	5.8	<0.1–109	3.1	1.9–19.4	5.8	5.6–27.1	16.5	22–69	47	21–69	47	28–40	32
Dean	—	2, 4, 7	O	7,700–9,100	8,200	6.0–68	26.2	<0.1–0.3	0.12	—	—	7.5–12.7	10.3	22–44	33.7	20–52	34	19–51	33	37–40	39
Delaware	—	1[c]	G	4,700–5,000	4,800	5.2–39	18.6	1.1–33	12.9	—	—	13.8–21.8	17.9	2.0–10.3	6.0	45–66	53	36–63	49	35–42	40
	—		O	3,500–5,100	4,200	3.0–52	14.5	0.6–84	24.5	—	—	15.2–25.4	21.0	3.9–15.6	11.2	33–65	49	31–64	42	—	—
Devonian	1	2, 5[d]	O	11,200–11,600	11,400	14–117	54	0.5–36	10.5	0.1–1.3	0.5	1.7–5.3	3.3	2.1–6.6	3.7	37–68	53	37–68	53	48–52	49
	2	2, 5[e]	G	11,300–12,300	11,800	8–299	99	0.2–23	4.0	0.1–5.8	0.8	1.3–6.8	4.3	3.3–16.7	9.2	19–53	33	19–53	33	—	—
			O	—	9,200	—	17	—	0.4	—	7.0	—	6.7	—	5.7	—	62	—	61	—	—
			O	5,500–9,900	7,700	8–113	34	2.5–50	14.9	0.2–18	7.0	5.5–27.7	15.2	6.8–22.9	11.0	41–76	51	41–76	51	35–46	42
			O	11,000–11,200	11,100	19–34	27	<0.1–2.2	1.1	<0.1–0.9	0.5	2.2–7.7	5.0	3.1–4.8	4.0	45–69	57	45–69	57	—	—
	3	2, 3, 4, 5, 6, 7, 8[f]	C	7,800–12,800	11,200	6.5–954	69	1.0–2,840	177	0.3–1,020	37	1.8–25.2	6.0	5.3–24.6	12.9	22–65	46	22–65	46	36–49	42

(Continued)

TABLE 2.4 (i) West Texas-Southeastern New Mexico Areas (Continued)

Formation	Group[a]	Range Area[b]	Fluid prod. depth prod.	Range of prod. depth, ft	Avg. prod. depth, ft	Range of prod. thickness, ft	Avg. prod. thickness, ft	Range of perm. K, md	Avg. perm. K, md	Range of perm. K90, md	Avg. perm. K90, md	Range of porosity, %	Avg. porosity, %	Range of oil satn, %	Ave. oil satn, %	Range of total water satn, %	Avg. total water satn, %	Range of connate water satn, %	Avg. connate water satn, %	Range of gravity, °API	Avg. gravity, °API
Ellenburger	—	All	C	4,100–10,600	7,400	11–18	14.3	203–246	225	1.4–54	27.7	3.7–4.6	4.2	0.8–7.6	4.2	47–67	57	47–67	57		
Fusselman	—	All	O	5,500–16,600	10,100	3.0–347	55	0.1–2,250	75	<0.1–396	22.9	1.3–13.8	3.8	1.0–19.2	8.4	40–84	61	40–84	60	37–52	47
Glorietta	—	All	C	8,700–12,700	10,300	18–51	34	1.2–26	8.4	0.3–1.3	0.9	2.6–3.7	3.3	0.2–3.9	1.7	32–47	40	32–47	40		
(Paddock)[g]	—	All	O	9,500–12,500	12,000	8–49	32	0.5–25	10.3	0.2–17	3.9	1.4–10.7	3.3	5.2–16	10.4	25–64	42	26–64	38	47–50	48
Granite wash	—	3,4,6,7	O	2,200–2,600	2,400	3–44	16.3	4.6–12	5.6		9.3	14–18.2	15.0	3.9–4.4	3.7	39–60	51	39–60	50	28–40	33
Grayburg			G	2,300–6,000	4,300	3–103	22.3	0.4–223	11.5	0.2–126	8.1	5.2–20.9	13.6	3.1–22.1	15.4	24–72	48	24–71	47		
			O	3,000–8,600	4,700	4–8	5.1	11–2,890	477		53	12.1–20.4	14.4	2.9–8.7	5.2	39–66	55	39–66	53	40–45	42
			G	2,300–3,400	3,000	2–81	15.6	5–3,290	609		30	3.5–26.1	17.7	4.8–22.5	14.7	42–71	54	35–66	49		
Grayburg	1	8	O	3,600–4,200	3,800	3.0–5.0	4.2	1.6–9.3	6.5			11.1–14.3	12.4	7.1–42	18.6	22–53	39	22–52	38	31–41	36
			O	2,400–4,500	4,100	3.0–123	27.4	0.5–159	13.7	0.2–48	5.2	7.0–20.0	11.3	6.2–37.9	17.6	26–56	36	25–55	35		
2		4,5,6,7	O	4,400	4,400	12–26	20.8	0.2–37	2.5	0.3–2.1	1.3	6.3–6.6	6.4	2.4–7.1	4.7	55–68	60	55–68	60	23–40	32
			O	3,000–4,800	4,400	6–259	45	0.2–118	5.5	0.1–110	2.7	2.7–16.2	7.9	4.8–22.1	13.9	32–84	55	32–84	55	28–35	31
3		1,2,3	O	1,300–3,900	2,700	4.5–182	50	0.3–1,430	37.7	0.1–228	14.3	5.3–24.3	11.9	8.3–34	18.2	31–78	58	31–78	56	38–47	41
Pennsylvanian sand (Morrow)[g]	—	2,3,4	O	4,100–11,400	9,100	1.7–77	22.3	0.3–462	34.9	0.1–168	14.7	2.7–13.9	7.7	4.7–18.8	9.7	28–58	42	28–58	41		
Queen (Penrose)[g]	—	All	G	3,000–3,200	3,100	4.0–29	9.9	10–318	64			10.7–22.2	16.6	2.6–7.6	7.4	36–62	48	35–58	45		
	—		O	800–4,900	3,500	1.5–38	10.2	0.2–4,190	123		1.0	5.7–27.0	17.2	4.2–34.7	15.6	32–68	49	30–66	45	30–42	33
San Andres	1	8	O	3,900–4,700	4,500	6–39	18.6	0.3–461	61	0.1–482	53	3.2–14.0	8.5	8.9–33.9	18.7	21–49	36	19–49	36	34–38	37
	2	5,6	O	4,100–5,300	4,500	4.7–124	40.1	0.3–295	6.9	0.1–208	3.8	3.1–12.8	7.1	4.9–30.6	14.7	26–69	52	25–69	51	30–37	33
	3	1,2,3,4,7	O	1,500–5,100	3,300	3.0–197	30.2	0.2–593	9.7	0.2–510	8.4	3.3–25.1	15.5	3.5–24.2	13.2	39–74	58	37–74	56	26–37	32

Seven Rivers	—	1, 2, 5, 8	G	3,600-4,100	3,900	3.0-8.0	5.6	0.6-23	12.2	—	0.3	15.5-16.6	16.0	3.4-9.5	6.4	51-66	56	46-65	54	28-38	32
Sprayberry	1	4,7	O	800-4,000	2,600	4.0-136	18.5	0.4-428	51.4	—	8.0	5.9-28.9	16.5	4.2-41.7	16.2	38-70	54	38-61	50	36-42	39
Strawn Lime	—	All	G	4,800-8,500	7,100	2.0-59	21.7	0.2-71	6.3	—	4.0	10.1-23.3	15.8	7.0-24.5	15.3	32-68	45	30-67	43		
			O		5,600	11-57	34.4	4.5-310	179	27-189	108	10.9-14.8	12.9	5.5-6.3	5.9	38-39	39	38-39	38	39-47	41
Strawn Sand	—	2, 3, 4, 7	G	5,200-6,700	5,900	2.0-101	36.7	1.9-196	43	0.6-148	19.1	31-12.6	7.2	4.9-28.3	11.2	48-60	52	15-66	52	29-48	42
			O	3,800-10,500	7,800	3.0-39	16.8	0.3-42	11.4	0.1-0.4	0.2	2.1-14.2	6.9	1.7-5.2	3.0	23-77	43	48-60	41		
		Others[h]	O	1,100-11,300	5,200	2.0-76	15.1	0.2-718	47	0.1-138	11.7	1.0-20.3	12.6	2.7-27.9	12.2	25-60	43	23-59	41	38	38
Tubb	—	1, 2, 5, 8	O	915-7,366	3,938	6-21	14	1.0-400	45	—	—	6.0-27	16.2	5.0-27	14.1	37-64	54	37-64	54		
	1	2	O	6,100-7,300	6,500	15-43	33.5	0.2-135	27.6	0.1-1.1	0.5	2.5-7.1	4.9	8.5-25.3	12.9						
Wolfcamp (Abo)[g]	2	2	O		9,800	—	10.6	—	2.3	—	0.4	—	4.3	—	19.6	—	25	—	25	42	42
	2	4, 5, 6, 7	O	8,400-9,200	8,800	13-129	41.7	2.3-9,410	419	1.5-6,210	274	4.9-18.5	9.9	6.6-16.8	9.7	32-68	44	31-56	44	36-45	40
	3	3 (part)	G	2,500-4,100	3,600	4.0-119	22.5	0.1-1,380	57	—	3.4	7.2-24.5	15.3	0.5-16.1	4.6	30-64	48	26-64	45	40-50	48
			O	2,400-4,100	3,500	2.0-114	28.0	1.0-1,270	60	—	4.3	5.4-26.3	15.5	1.6-26.8	14.3	32-65	46	29-64	44	40-44	42
	4	8	O	9,000-10,600	9,700	4.5-204	59	0.2-147	20.4	0.2-36	5.4	2.6-12.8	8.1	5.3-23.6	14.4	28-56	39	28-56	39		
Yates	—	1, 2, 5, 8	G	1,400-3,500	2,800	3.0-53	10.8	0.2-145	19.3	—	—	12.1-27.4	17.9	1.3-17.0	5.6	43-79	59	36-78	53	27-41	32
			O	1,400-4,000	2,300	3.0-66	16.6	1.0-4,000	42.7	—	27.8	2.4-27.0	18.8	3.7-37.3	16.0	31-75	53	31-75	47		

[a] More than one group indicates distinct differences in formation as found in different areas.

[b] Area numbers refer to map of Figure 24.1.

[c] Plus Ward, Pecos, and Southwestern Lea County.

[d] Midland and Ector counties only.

[e] Crane, Ward, Winkler, and Pecos counties only.

[f] Except counties in 4 and 5 above.

[g] Commonly used in New Mexico, Area 8.

[h] Archer, Baylor, Clay, Jack, Montague, Wichita, Wise, and Young counties.

TABLE 2.5   Particle Size Definitions

Material	Particle size, μm	U.S. standard sieve mesh no.
Coarse sand	>500	<35
Medium sand	250–500	35–60
Fine sand	125–250	60–120
Very fine sand	62.5–125	120–130
Coarse silt	31–62.5	—
Medium silt	15.6–31	—
Fine silt	7.8–15.6	—
Very fine silt	3.9–7.8	—
Clay	<2	—

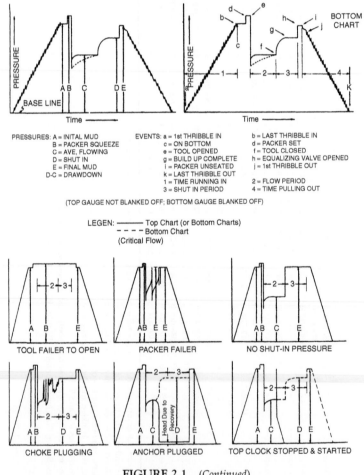

FIGURE 2.1   (*Continued*)

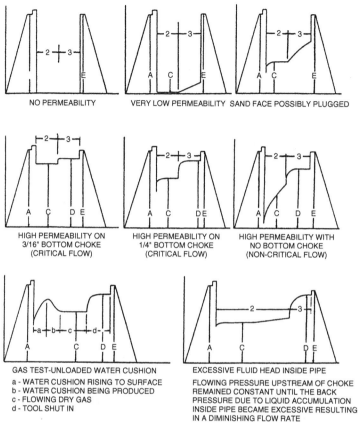

NO PERMEABILITY   VERY LOW PERMEABILITY SAND FACE POSSIBLY PLUGGED

HIGH PERMEABILITY ON
3/16" BOTTOM CHOKE
(CRITICAL FLOW)

HIGH PERMEABILITY ON
1/4" BOTTOM CHOKE
(CRITICAL FLOW)

HIGH PERMEABILITY WITH
NO BOTTOM CHOKE
(NON-CRITICAL FLOW)

GAS TEST-UNLOADED WATER CUSHION

a - WATER CUSHION RISING TO SURFACE
b - WATER CUSHION BEING PRODUCED
c - FLOWING DRY GAS
d - TOOL SHUT IN

EXCESSIVE FLUID HEAD INSIDE PIPE

FLOWING PRESSURE UPSTREAM OF CHOKE
REMAINED CONSTANT UNTIL THE BACK
PRESSURE DUE TO LIQUID ACCUMULATION
INSIDE PIPE BECAME EXCESSIVE RESULTING
IN A DIMINISHING FLOW RATE

*NOTE: THE RUNNING IN AND PULLING OUT PERIODS IN THOSE CHARTS
ARE SHOWN COMPRESSED ON TIME SCALE FOR CLARITY.*

FIGURE 2.1

2.3 LOGGING

2.3.1 Introduction

This section deals with the part of formation evaluation known as well logging. Well logs are a record versus depth of some physical parameter of the formation. Parameters such as electrical resistance, naturally occurring radioactivity, or hydrogen content may be measured so that important producing characteristics such as porosity, water saturation, pay thickness, and lithology may be determined. Logging instruments (called *sondes*) are lowered down the borehole on armored electrical cable (called a *wireline*). Readings are taken while the tool is being raised up the hole. The information is transmitted uphole via the cable where it is processed by an on-board

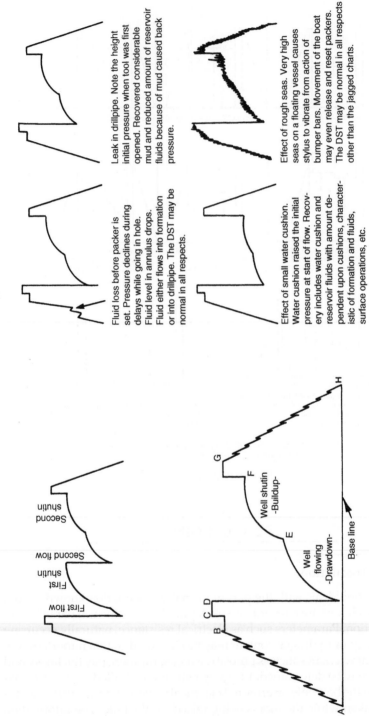

Leak in drillpipe. Note the height initial pressure when tool was first opened. Recovered considerable mud and reduced amount of reservoir fluids because of mud caused back pressure.

Effect of rough seas. Very high seas on a floating vessel causes stylus to vibrate from action of bumper bars. Movement of the boat may even release and reset packers. The DST may be normal in all respects other than the jagged charts.

Fluid loss before packer is set. Pressure declines during delays while going in hole. Fluid level in annulus drops. Fluid either flows into formation or into drillpipe. The DST may be normal in all respects.

Effect of small water cushion. Water cushion raised the initial pressure at start of flow. Recovery includes water cushion and reservoir fluids with amount dependent upon cushions, characteristic of formation and fluids, surface operations, etc.

First flow
First shutin
Second flow
Second shutin

A
B
C D
Well flowing -Drawdown-
E
Well shutin -Buildup-
F
G
H
Base line

FIGURE 2.2

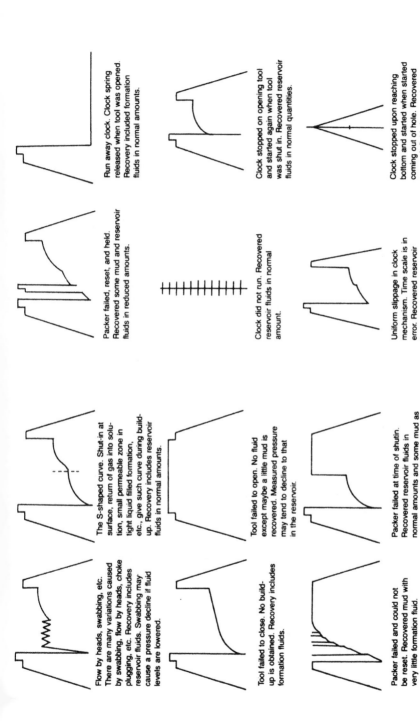

Flow by heads, swabbing, etc. There are many variations caused by swabbing, flow by heads, choke plugging, etc. Recovery includes reservoir fluids. Swabbing may cause a pressure decline if fluid levels are lowered.

Tool failed to close. No build-up is obtained. Recovery includes formation fluids.

Packer failed and could not be reset. Recovered mud with very little formation fluid.

The S-shaped curve. Shut-in at surface, return of gas into solution, small permeable zone in tight liquid filled formation, etc., give such curve during build-up. Recovery includes reservoir fluids in normal amounts.

Tool failed to open. No fluid except maybe a little mud is recovered. Measured pressure may tend to decline to that in the reservoir.

Packer failed at time of shutin. Recovered reservoir fluids in normal amounts and some mud as required to equalize pressure.

Packer failed, reset, and held. Recovered some mud and reservoir fluids in reduced amounts.

Clock did not run. Recovered reservoir fluids in normal amount.

Uniform slippage in clock mechanism. Time scale is in error. Recovered reservoir fluids in normal amounts.

Run away clock. Clock spring released when tool was opened. Recovery included formation fluids in normal amounts.

Clock stopped on opening tool and started again when tool was shut in. Recovered reservoir fluids in normal quantities.

Clock stopped upon reaching bottom and started when started coming out of hole. Recovered reservoir fluids in normal quantities.

FIGURE 2.2 (Continued)

Clock stopped at shut-in. Drillstem normal. Recovered reservoir fluids in normal quantities.

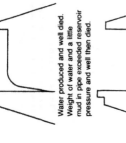

Stylus dragging. Pressure chart has stair step character. Stylus needs adjustment. Drillstem test normal in all respects.

Water produced and well died. Weight of water and a little mud in pipe exceeded reservoir pressure and well then died.

Effect of large super-pressure. Pressure build-up during flow and build-up period is more rapid than usual. Mud recovery also is probably greater. Highest pressure may or may not exceed normal reservoir pressure.

Re-solution of gas in drillpipe when well is shut-in at surface. Test is probably normal in all respects.

Flow string plugs immediately above gauge. Recovered little fluid, mostly mud. Pressure builds quickly to that of reservoir.

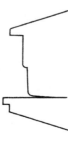

Low formation permeability. Recovered very little mud and some formation fluid.

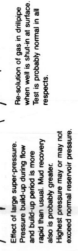

Plugging and unplugging of flow stream at some location above the gauge. Recovered reservoir fluids, possibly in reduced amounts.

Gradual plugging of flow stream below gauge. Pressure declines as flow rate decreases to a value equal to weight of fluids above gauge. Recovered a little mud and small amount of reservoir fluids.

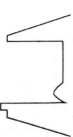

No formation permeability. Small amount of mud may be recovered with very little formation fluid.

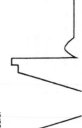

Effect of skin. Recovered formation fluids. The amount recovered, pressure increase during flow, rate of build-up depends upon permeability, the pressures and differentials, compressibility of fluids, volumes affected, and degree of damage from solid and filtrate invasion, perforations, partial penetration of the formation, etc. A large skin can reduce flow substantially and very large pressure differentials can result from skins.

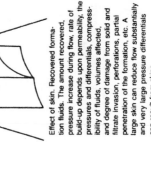

Flow on choke from highly permeable formation. The surface and subsurface flowing pressures and flow rate depend upon transmissibility, formation pressure, pressure differentials, choke size and many other characteristics of the formation, fluids, and the flow system. Build-up during flow and shut-in is very rapid and reliable values for skin and permeability may not be obtainable.

FIGURE 2.2

Unloading water cushion in gas well.
a. Water cushion rising to surface
b. Water cushion being produced
c. Flowing dry gas with choke

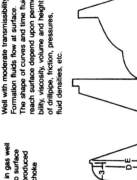

Excessive fluid head inside pipe. Flowing pressure upstream of the surface or subsurface choke remained constant until back pressure due to liquid accumulation became large, resulting in a lower flow rate.

Change in size of pipe string. The change in slope during flow may be in either direction depending upon where larger pipe is located in string. Normal recovery of reservoir fluids.

Well with moderate transmissibility. Formation fluids flow at surface. The shape of curves and time fluids reach surface depend upon permeability, viscosity, volume and height of drillpipe, friction, pressures, fluid densities, etc.

Two layer effects. Caused by two producing zones, poorly connected lenses, fault system, discontinuity, fluid boundaries, etc. Curves change slope — often during flow and build-up and such change may be in two directions depending upon conditions.

Gauge plugged while going into hole. Plugging occurred when fluid weight equalled recorded pressure and unplugged at lower pressure when coming out of hole. Normal recovery of reservoir fluids.

Gauge plugged with tool on bottom before packer was set. Unplugged at same pressure when coming out of hole. Reservoir fluids recovered in normal amount.

Gauge gradually plugged during flow period. Unplugged when coming out of hole. Reservoir fluids recovered in normal quantities.

Hole in chart

Stylus tore chart and could not move further. Reservoir fluids recovered in normal amounts.

Gauge plugged after packer was set but before tool was opened. Unplugged at lower pressure while coming out of hole. Reservoir fluids recovered in normal amounts.

Gauge plugged during flow and unplugged late in build-up. Reservoir fluids recovered in normal amounts.

Well interference. The DST is usually too short to notice interference with today's well spacing. Pressure declines at long times. Reservoir fluids may be recovered in about normal quantities.

FIGURE 2.2 (Continued)

Two tests with the same gauge. Second build-up extrapolated pressure is lower than that of first build-up. Depletion of a small reservoir might be suggested.

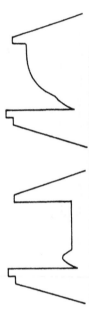

Two tests with the same gauge. Character of the curves for the second test differs appreciably from that of the first test. Skin or other parameter in the flow and build-up equations which might be sensitive to flow or shutin has changed between the two tests. Since some fluid entered drillstring during first test, initial pressure of second test will be higher than that of first test by the weight of this fluid column.

Single test with two gauges. The two sets of curves should be identical except for the small difference in pressure related to location of the gauges within the flow string. This is the normal condition to be expected if both gauges are operating properly and no plugging, etc., is occurring.

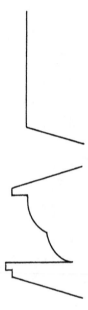

Single test with two gauges. Left gauge suggests highly permeable formation with little or no skin — maybe negative — while right curve indicates gauge which plugged upon reaching bottom and unplugged when starting out of hole. If right gauge was at bottom, it was probably stuck in debris at bottom of hole.

Single test with two gauges. Left gauge shows gradual plugging below gauge while right gauge shows actual behavior of the reservoir. Left gauge measures weight of fluids above gauge rather than reservoir characteristics. The two are not identical because a plug in the flow string has removed the connection between the two pressure sources which are in balance in the normal test.

Single test with two gauges. Left gauge measures reservoir characteristics while right gauge plugged while going into the well and did not unplug until inspected at surface.

FIGURE 2.2

computer and recorded on magnetic tape and photographic film. In older logging units, downhole signals are processed by analog circuits before being recorded.

Well logging can be divided into two areas: open hole and cased hole. Open hole logging is done after drilling, before casing is set. The purpose of open hole logging is to evaluate all strata penetrated for the presence of oil and gas. Open hole logs give more reliable information on producing characteristics than cased hole logs. Cased hole logs provide information about cement job quality, casing corrosion, fluid flow characteristics, and reservoir performance. In areas where the geologic and producing characteristics of a reservoir are well known, as in development wells, cased hole logs are used for correlation. In recent years, many new open- and cased-hole logs and services have become available, including fluid samplers, sidewall cores, fracture height log, and seismic services. These products, in conjunction with new computer processing techniques, provide the engineer and geologist with an enormous amount of data for any well.

2.3.2 Parameters that Can Be Calculated or Estimated from Logs

2.3.2.1 Porosity

Porosity is defined as the ratio of volume of pores to the to total volume of the rock. It occurs as primary (depositional) or secondary (diagenetic or solution) porosity. Primary and secondary porosity can be read directly from neutron, density, and sonic logs. These tools do not measure pore volume directly, rather they measure physical parameters of the formation and relate them to porosity mathematically or empirically. Since the sonic tool only records primary (or matrix) porosity, it can be combined with total porosity tools, such as density or a combined neutron and density, to determine secondary porosity:

$$\phi_{secondary} = \phi_{total} - \phi_{sonic} \qquad (2.4)$$

where $\phi_{secondary}$ = porosity due to vugs and fractures
ϕ_{total} = total porosity as determined from cores, density log, neutron-density crossplot, or localknowledge
ϕ_{sonic} = porosity determined from sonic log.

The combinable magnetic resonance (CMR) tool capable of measuring porosity independent of lithology and measuring movable or bound fluid, which can be used to measure effective porosity in the rock.

2.3.2.2 Water Saturation

Connate water saturation (S_w) and flushed zone water saturation (S_{xo}) can be calculated from information supplied by well logs.

Connate water saturation is the fraction of pore volume in an undisturbed formation filled with connate water.

$$S_W = \frac{\text{volume of water}}{\text{volume of pores}} \qquad (2.5)$$

Flushed zone saturation (S_{xo}) is the fraction of the pore volume filled with flushing agent (normally drilling fluid).

$$S_{xo} = \frac{\text{volume of flushing agent}}{\text{volume of pores}} \qquad (2.6)$$

Prior to penetration by a drill bit, only two fluids are assumed to be present in the formation-water and hydrocarbons. Therefore, all pore space that is not occupied by water is occupied by hydrocarbons. With this assumption hydrocarbon saturation can be calculated:

$$S_h = 1 - S_W \qquad (2.7)$$

where S_h = hydrocarbon saturation.

Pay Thickness   The thickness of a hydrocarbon-bearing formation (h_{pay}) is easily determined from well logs once ϕ and S_W cutoffs are established. The S_W cutoff is the maximum value for S_W for a given rock type. The ϕ cutoff is the minimum value for ϕ below which hydrocarbons cannot be produced. For example:

Depth	ϕ	S_W	Comment
3,668–3,670	1%	53%	ϕ too low
3,666–3,668	2%	50%	ϕ too low
3,664–3,666	6%	38%	possible hydrocarbons
3,662–3,664	6%	36%	possible hydrocarbons
3,660–3,662	8%	31%	possible hydrocarbons
3,658–3,660	7%	74%	too wet
3,656–3,658	8%	100%	too wet

In this case, the water saturation cutoff is a maximum of 60% and the porosity cutoff is a minimum of 3%, so this well will have 6 feet of pay ($h_{pay} = 6\,\text{ft}$). Other factors that may reduce h_{pay} include shaliness, shale streaks, low permeability, and low reservoir pressure. Porosity and water saturation cutoffs are usually established for specific regions or reservoirs based on detailed production and geologic information.

2.3.2.3 Lithology

It is often necessary to know the rock type in order to properly design downhole assemblies, casing programs, and completion techniques. Data from well logs can provide the geologist or engineer with an estimate of the lithologic makeup of any formation. The accuracy of this estimate is a function of the complexity of the formation (mineralogic makeup and fluid types) and the kinds of tools used to investigate the rocks. More tools are needed to accurately determine compositions of complex formations. Simple lithologies (three or less minerals, or gas) can be determined with combined neutron, density, and sonic logs. This technique will be discussed later. More complex lithologies can be determined with the aid of special logging tools and computers.

Since well logs infer lithology from physical and chemical parameters, certain rocks will look the same on logs though they differ in their geologic classification. Sandstone, quartz, and chert are all SiO_2 and appear the same on porosity logs. The same is true of limestone and chalk. Dolomite, anhydrite, and salt have very distinct characteristics and are easily distinguished from other rock types. Shales are composed of clay minerals. The type and amount of different clay minerals, which vary widely between shales, can affect their bulk properties.

2.3.2.4 *Permeability*

Permeability is one of the essential properties used in evaluation of a potentially producing formation. Unfortunately there are no logging devices that read permeability. This is because permeability is a dynamic property. Most logging tools spend only a few seconds in front of any one point of a formation, therefore it is impossible to measure any time-dependent parameter. There are methods to estimate permeability from well logs, but they are based on general assumptions. From a practical standpoint, log parameters only provide an "order of magnitude" approximation. Several methods of inferring permeability with well logs are discussed where applicable in each section.

Two relationships between porosity and irreducible water saturation (S_{wi}) are used to estimate permeability:

1. The Timur relationship [19] (Figure 2.3) for granular rocks (sandstones and oolitic limestones), which generally gives a more conservative estimate of permeability.
2. The Wyllie and Rose relationship [43] modified by Schlumberger [20] (Figure 2.4), which generally gives a higher estimate of permeability.

To enter these charts, porosity and irreducible water saturation (S_{wi}) must be known. Porosity can be obtained from cores or any porosity device (sonic, neutron, or density). Irreducible water saturation must be found from capillary pressure curves or it can be estimated. The permeabilities from these charts should be considered "order of magnitude" estimates.

2.3.3 Influences on Logs

The purpose of well logging is to determine what fluids are in the formation and in what quantity. Unfortunately the drilling process alters the

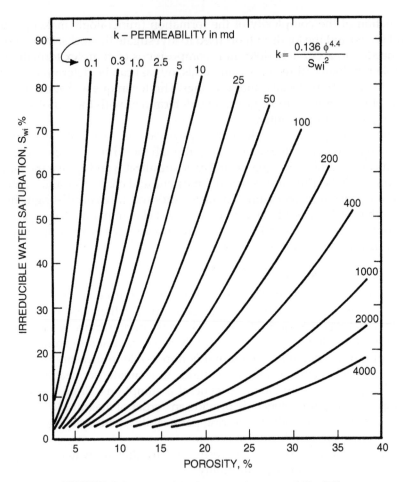

FIGURE 2.3 Timur chart for estimating permeability [19].

fluid saturations by flushing the pores near the borehole and filling them with the fluid fraction of the drilling mud (mud filtrate). To correct for these influences, the invasion profile must be identified. Figure 2.5, an idealized cross-sectional view of the borehole and formations, shows an invasion profile and the appropriate symbols for each part of that profile [20].

Mud Relationships   Since the borehole is filled with mud and the adjacent portion of the formation is invaded with mud filtrate, mud properties must be accurately known so they can be taken into account. Mud has a minor influence on most porosity tools; however, it can have a large effect on the resistivity tools. In general:

$$R_{mc} > R_m > R_{mf} \tag{2.8}$$

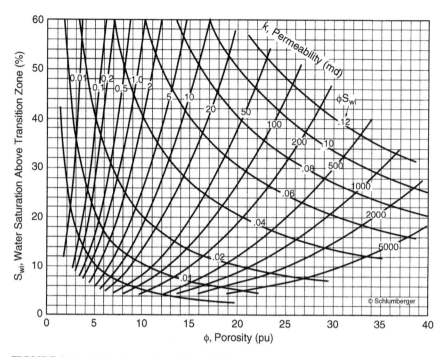

FIGURE 2.4 Schlumberger chart (after Wyllie and Rose) for estimating permeability [20].

This is because the mudcake is mostly clay particles and has very little water associated with it. The clay particles in the mudcake tend to align themselves parallel to the borehole wall, developing a high horizontal resistivity R_h (Figure 2.6). Since the mud filtrate is composed only of fluid and has no solids, it will have a lower resistivity. If the resistivity of the mud (R_m) is known, R_{mc} and R_{mf} can be estimated with the following equations:

$$R_{mf} = K_m(R_m)^{1.07} \qquad (2.9)$$

$$R_{mc} = 0.69(R_{mf})\left(\frac{R_m}{R_{mf}}\right)^{2.65} \qquad (2.10)$$

where $K_m = 0.847$ for   10 lb/gal
           0.708   11 lb/gal
           0.584   12 lb/gal
           0.488   13 lb/gal
           0.412   14 lb/gal
           0.380   16 lb/gal
           0.350   18 lb/gal

These relationships work well for most muds (except lignosulfonate) with resistivities between 0.1 Ω-m and 10 Ω-m.

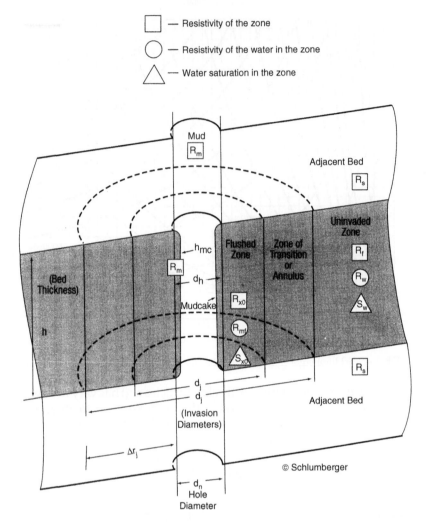

FIGURE 2.5 Diagram of the borehole environment showing the various zones and their parameters [20].

Another approximation that works well for salt muds is:

$$R_{mf} = 0.75\,R_m \tag{2.11}$$

$$R_{mf} = 1.5\,R_m \tag{2.12}$$

Temperature Relationships   Mud resistivity is a function of temperature and ion concentration. Since temperature increases with depth due to geothermal gradient, the mud resistivity is lower at the bottom of the hole than at the surface (pits). The temperature of a formation can be found with

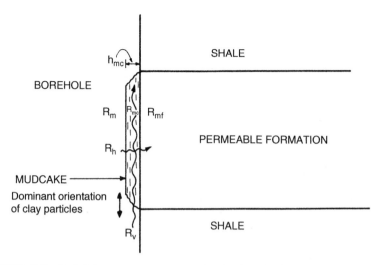

FIGURE 2.6 Resistivity components on mud cake that develop opposite a permeable formation.

an equation suggested by Hilchie [21]:

$$T_f = (T_{TD} - T_s)\frac{D_f}{D_{TD}} + T_s \qquad (2.13)$$

where T_f = formation temperature (°F)
T_{TD} = temperature at total depth (BHT) (°F)
T_s = average surface temperature (°F)
D_f = formation depth (ft)
D_{TD} = total depth (ft)

Average surface temperatures in various oilfield areas are:

Alberta	40°F
California	65°F
Colorado-Northern New Mexico	55°F
Gulf Coast	80°F
Oklahoma	65°F
Permian Basin	65°F
Wyoming	45°F

Figure 2.7 solves this equation graphically [20].

Fluid resistivity at any formation depth can be found using the Arps Equation if the resistivity at any temperature and formation temperature are known [20]:

$$R_2 = R_1\frac{T_1 + 6.77}{T_2 + 6.77} \qquad (2.14)$$

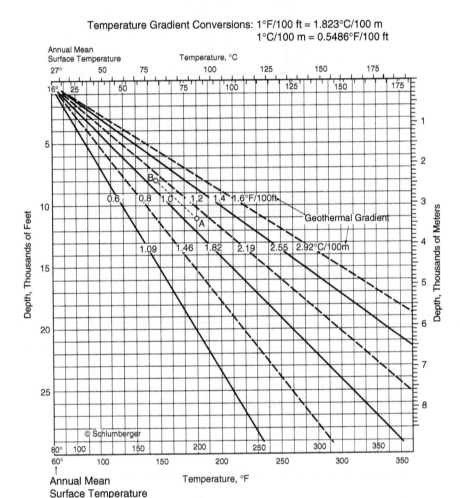

EXAMPLE: Bottom hole temperature, BHT, is 200°F at 11,000 ft (Point A).
Temperature at 8,000 ft is 167°F (Point B).

FIGURE 2.7   Chart for estimating formation temperature [20].

where $R_2 =$ fluid resistivity at formation temperature, $\Omega - m$
$\quad\quad R_1 =$ fluid resistivity at some temperature, $\Omega - m$
$\quad\quad T_2 =$ formation temperature, °F
$\quad\quad T_1 =$ temperature R_1 was measured at, °F

A monograph that solves this equation graphically is presented in Figure 2.8 [20].

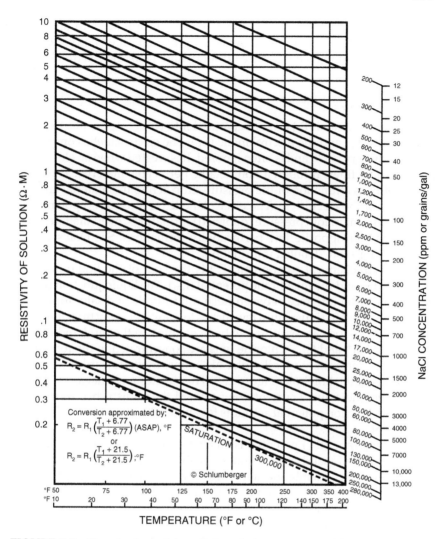

FIGURE 2.8 Chart for determining salinity, solution resistivity and for converting resistivity to formation temperature [20].

2.3.4 Openhole Logs and Interpretation

SP (Spontaneous Potential) The SP log has 4 basic uses: (1) recognition of permeable zones, (2) correlation of beds, (3) determination of R_w, and (4) qualitative indication of shaliness. The SP can only be used in fresh mud and is run with several resistivity tools. The curve is presented in Track 1 and is scaled in millivolts (mV).

The SP log is a record of the naturally occurring electrical currents created in the borehole. These currents or circuits usually occur at bed boundaries and are created by the interaction between fresh drilling mud and salty formation water. The curve represents the potential difference between a stationary electrode on the surface (ground) and a moving electrode in the borehole.

2.3.4.1 Theory

The total potential (E_t) can be separated into two components: the electrochemical (E_c) and the electrokinetic (E_k). The electrokinetic component is generally very small and is often ignored. It is created when an electrolyte (mud filtrate) flows through a nonmetallic permeable material. The magnitude of E_k is a function of the pressure drop across the material and the resistivity of the electrolyte. The electrokinetic (or streaming) potential is most significant in low pressure (depleted) formations, overbalanced mud conditions and opposite low permeability formations. See Doll's classic paper [22] for more detailed information.

The electrochemical component (E_c) is the sum of the liquid-junction potential (E_{lj}) and the membrane potential (E_m). The liquid junction potential occurs at the interface between fresh mud filtrate and salty formation water. This interface is usually a few inches to a few feet away from the borehole. Only two ions are assumed to be in solution in the mud and formation water: Na^+ and Cl^-. Chloride ions are concentrated in the formation water, and being more mobile than Na^+ ions, move toward lower concentrations in the borehole (Figure 2.9). This creates a net negative charge near the borehole and a current flows toward the undisturbed formation. The liquid-junction potential accounts for about 20% of the electrochemical component.

The membrane (E_m) potential is created at the bed boundary between a permeable bed (sand) and an impermeable bed (shale). The shale acts as an ion-selective membrane, allowing only the smaller Na^+ ion to move through the clay crystal structure from the salty formation water toward the fresh drilling fluid in the bore. This creates a net positive charge along the shale. It also creates a large concentration of negative charges associated with the Cl^- ion in the permeable bed. This phenomena is also shown in Figure 2.9. The membrane potential accounts for about 80% of the electrochemical potential. The total effect of these two potentials is a net negative charge within the permeable zone when the connate water is saltier than the mud filtrate.

2.3.4.2 Interpretation

The total electrochemical component of the total potential is what the SP records. It can be calculated with the following equation:

$$E_c = -K \log \frac{a_w}{a_{mf}} \qquad (2.15)$$

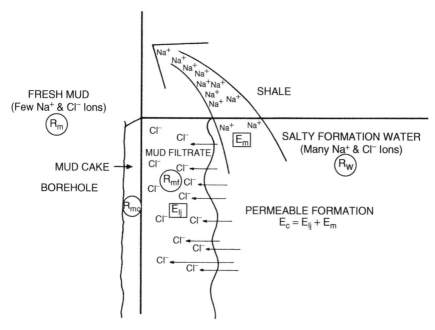

FIGURE 2.9 Ionic movement that contributes to the development of an SP curve.

where $-K = -(0.133T + 61)$ (T in °F)
a_w = chemical activity of formation water
a_{mf} = chemical activity of drilling mud filtrate

Since the chemical activity of a solution cannot be used, it must be converted to its equivalent electrical resistivity. Chemical activity of a fluid is approximately equal to the inverse of its equivalent electrical resistivity. Conversion to equivalent resistivities makes the equation:

$$E_c = -K \log \frac{R_{mfeq}}{R_{weq}} \qquad (2.16)$$

Since E_c is equal to the maximum SP deflection recorded on a log (SSP), Equation 2.16 can be rewritten to read:

$$SSP = -K \log \frac{R_{mfeq}}{R_{weq}} \qquad (2.17)$$

where SSP is the static (or maximum) spontaneous potential recorded opposite a permeable formation. Since the purpose of an SP log is to find R_{weq} and then R_w, if we know SSP we can solve Equation 2.17 for R_{weq}:

$$R_{weq} = \frac{R_{mfeq}}{10^{(-SSP/-K)}} \qquad (2.18)$$

Once R_{weq} is known, it is converted to R_w using the chart shown in Figure 2.10 [20]. SSP can come directly from the log if the bed is thick and the SP curve reaches a constant value and develops a "flat top." If the curve is pointed or rounded, it must be corrected for bed thickness.

The shape and amplitude of the SP are affected by:

1. Thickness and resistivity of the permeable bed (R_t).
2. Diameter of invasion and resistivity of flushed zone (R_{xo})
3. Resistivity of the adjacent shales (R_s).
4. Resistivity of the mud (R_m).
5. Borehole diameter (d_h)

All of these must be accounted for when examining the SP, and any necessary corrections should be made. To find the magnitude of the SP, take

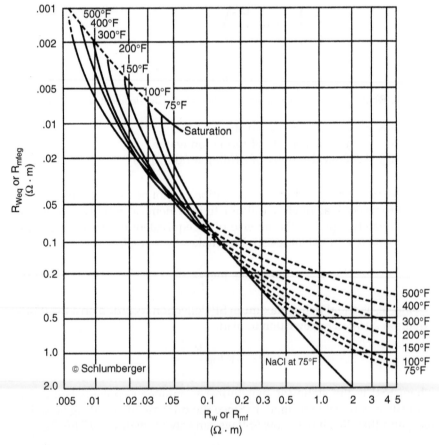

FIGURE 2.10 Chart for converting R_{mf} to R_{mfeq} and R_{weq} to R_w for SP-R_w calculations [20].

the maximum deflection from the average shale value (shale baseline) to the most negative value. (Figure 2.11 shows a curve that needs correction and one that does not.) Bed thickness corrections can be made from Figure 2.12 and should *always* increase the magnitude of the SP.

Another use for the SP log is finding permeable zones. Any negative deflection of the cure indicates a potentially permeable zone. The magnitude of the deflection has no relation to the amount of permeability (in

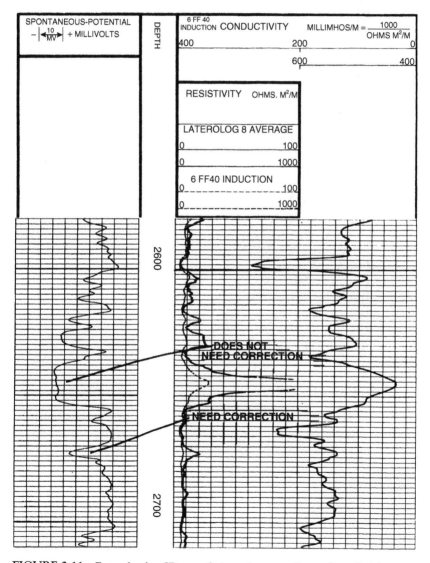

FIGURE 2.11 Example of an SP curve that requires correction and one that does not.

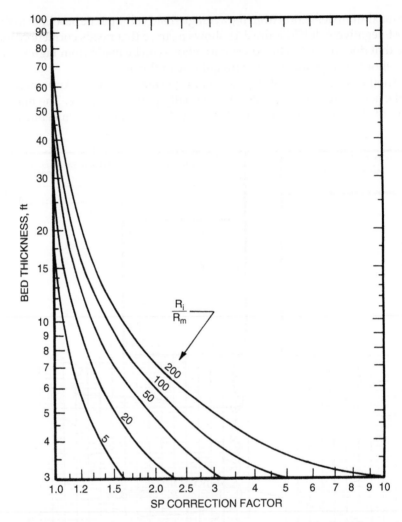

FIGURE 2.12   SP bed thickness correction chart for the SP [21, 22].

millidarcies); it merely indicates that the rock has ionic permeability. No quantitative information on this, parameter can be derived from the SP. Figure 2.13 shows an example of permeable and impermeable zones on an SP log.

Interpretation of an SP log follows a few basic rules:

1. If the SP curve is concave to the shale line, the formation is permeable.
2. If the SP curve is convex to the shale line, the formation is impermeable.
3. Constant slope means high resistivity–usually impermeable.
4. High resistivity formations cause the bed boundaries to become rounded.

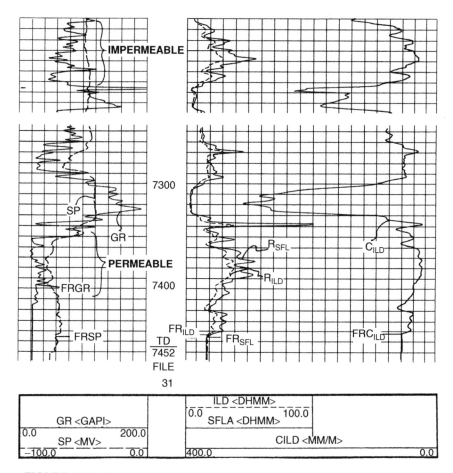

FIGURE 2.13 Examples of an SP log through an impermeable and a permeable zone.

5. A thin permeable bed does not reach maximum deflection.
6. A thin shale streak does not reach the shale baseline.
7. Bed boundaries are picked at the inflection points in clean sands. Bed boundaries should be confirmed with some other log such as the gamma ray.

Resistivity Tools The purpose of resistivity tools is to determine the electrical resistance of the formation (rock and fluid). Since most formation waters contain dissolved salts, they generally have low resistivities. Hydrocarbons do not conduct electricity, therefore rocks that contain oil and/or gas show high resistivity. This is the way hydrocarbon-bearing zones are differentiated from water zones.

Resistivity tools are divided into three types based on the way measurements are made: (1) non-focused (normal) tools, (2) induction tools, and (3) focused resistivity tools. Microresistivity tools will be treated under a separate heading.

Resistivity tools are further divided by depth of investigation. Tools may read the flushed zone (R_{xo}), the transition zone between the flushed and uninvaded zones (R_i), or the univaded zone (R_t). For purposes of this discussion, all tool names are for Schlumberger equipment. Comparable tools offered by other logging companies are summarized in Table 2.6.

2.3.4.3 Theory Nonfocused (Normal Tools)

The first tools to be used were nonfocused tools. The electrode- arrangement is as shown in Figure 2.14a. As shown earlier, resistivity, R, is found using Ohm's Law ($r = V/I$):

$$R = \frac{rA}{L} \qquad (2.44)$$

where V = voltage read from meter
     I = current read from meter
     A = surface area, m^2
     L = length, m
     R = resistivity, ohm-m
     r = resistance, ohms

The tools work well in low-resistivity formations with thick beds and in slightly conductive muds.

2.3.4.4 Inductions Tools

Since a slightly conductive mud is necessary for the normal tools, they cannot be used in very fresh muds or in oil-base muds. The induction tool overcomes these problems by inducing a current into the formation instead of passing it through the mud-filled borehole. Figure 2.14b shows a simplified two-coil induction tool. High-frequency alternating current is sent into the transmitter coil. The alternating magnetic field that is created induces secondary in circular paths called ground loops. These currents in turn create a magnetic field which induces currents in the receiver coil. The signal received is proportional to the formation conductivity. Conductivity readings are then converted to resistivity. Additional coils are used to focus the tool so that conductive beds as thin as four feet can be detected. Inductions tools work well in oil-base, foam, air, gas, and fresh mud. The induction tool is unreliable above 500 Ω-m and useless above 1,000 Ω-m or in salt muds. Readings are only considered reasonable below 100 Ω-m and are accurate between 1 and 20 Ω-m when mud is very fresh.

TABLE 2.6 Service Company Nomenclature

Schlumberger	Gearhart	Dresser Atlas	Welex
Electrical log (ES)	Electric log	Electrolog	Electric log
Induction electric log	Induction electrical log	Induction electrolog	Induction electric log
Induction spherically focused log			
Dual induction spherically focused log	Dual induction-laterolog	Dual induction focused log	Dual induction log
Laterolog-3	Laterolog-3	Focused log	Guard log
Dual laterolog	Dual laterolog	Dual laterolog	Dual guard log
Microlog	Micro-electrical log	Minilog	Contact log
Microlaterolog	Microlaterolog	Microlaterolog	$F_o R_{xo}$ log
Proximity log		Proximity log	
Microspherically focused log			
Borehole compensated sonic log	Borehole compensated sonic	Borehole compensated acoustilog	Acoustic velocity log
Long spaced sonic log		Long spacing BHC acoustilog	
Cement bond/variable density log	Sonic cement bond system	Acoustic cement bond log	Microseismogram
Gamma ray neutron	Gamma ray neutron	Gamma ray neutron	Gamma ray neutron
Sidewall neutron porosity log	Sidewall neutron porosity log	Sidewall epithermal neutron log	Sidewall neutron log
Compensated neutron log	Compensated neutron log	Compensated neutron log	Dual spaced neutron
Thermal neutron decay time log		Neutron lifetime log	Thermal multigate decay
Dual spacing TDT		Dual detector neutron	
Formation density log	Compensated density log	Compensated densilog	Density log
Litho-density log			
High resolution dipmeter	Four-electrode dipmeter	Diplog	Diplog
Formation interval tester	Selective formation tester	Formation tester	Formation tester
Repeat formation tester		Formation multitester	Multiset tester
Sidewall sampler	Sidewall core gun	Corgun	Sidewall coring
Electromagnetic propagation log			
Borehole geometry tool	X-Y caliper	Caliper log	Caliper
Ultra long spacing electric log			Compensated
Natural gamma ray spectrometry		Spectralog	Spectral natural gamma
Gamma ray spectroscopy tool log		Carbon/oxygen log	
Well seismic tool			
Fracture identification log	Fracture detection log		

From Reference 36.

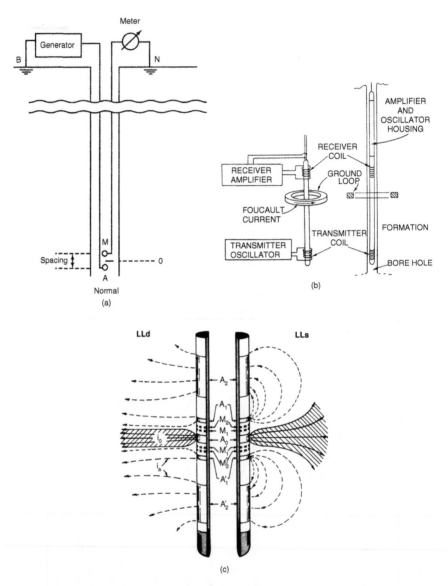

FIGURE 2.14   (a) Basic electrode arrangement of a normal device used in conventional electric logs. (b) Electrode arrangement for a basic, two coil induction log system. (c) Electrode arrangement and current flow paths for a dual laterolog sonde [52].

2.3.4.5 *Phasor Induction Tools*

Since the early 1960's, induction logging tools have become the principal logging device for fresh, slightly conductive to non-conductive (oil-base) muds. However, these devices are significantly affected by environmental (bore-hole size and mud composition) and geological (bed thickness and invasion) conditions. Also, high formation resistivities (> 50 to 100 Ω-m) dramatically increase the difference between apparent Rt and true Rt [20, 24]. In 1986, Schlumberger introduced a new induction log to off-set these problems [24]. This device is known by the trade name Phasor Induction SFL and uses a standard dual induction tool array. The difference between the conventional and phasor devices is in signal processing made possible by miniaturization of computer components. Induction tools all produce two signals: the inphase (R-signal) induction measurement and the quadrature (X-signal) induction signal. The R-signal is what is presented on standard dual induction–SFL log presentations. The R-signal and X-signal measurements are combined during advanced processing in the logging tool itself to produce a log with real time corrections for environmental and geological conditions. Apparent Rt is nearly equal to true Rt in most situations. Vertical resolution of this device is about the same as conventional induction tools (about 6 feet), but enhanced and very enhanced resolution phasor tools are available that have vertical resolutions down to 2 feet [24]. The primary advantages of this tool include much better Rt readings in high resistivity formations (i.e. > 100 Ω-m) and more accurate readings in salty muds than previously possible.

2.3.4.6 Focused Resistivity Tools (Laterologs)

The Laterologs are the primary salt-mud resistivity tools. Salt mud presents a problem in that the path of least resistance is within the borehole. Therefore the current must be forced into the formation which has higher resistance. To do this, secondary electrodes (A_1, and A_2) are placed above and below the measuring current-emitting electrode. (A_0). These secondary electrodes emit "focusing" or "guard" currents with the same polarity as the measuring currents. Small monitoring electrodes (M_1, M_2, M'_1 and M'_2) adjust the focusing currents so that they are at the same potential as A_1 and A_2. A sheet of current measuring one to two feet thick is then forced into the formation from A_0. The potential is then measured between M_2 and M_1 and a surface ground. Since I_o is a constant, any variation in M_1 and M_2 current is proportional to formation resistivity. Figure 2.14c shows the electrode arrangement [52].

2.3.4.7 Corrections

As previously mentioned, resistivity tools are affected by the borehole, bed thickness, and invaded (flushed) zone. If bed thickness and mud resistivity are known, these effects can be accounted for. Major service companies (Schlumberger, Atlas Wireline, Welex Halliburton) provide correction charts for their tools. Several charts for Schlumberger tools are included in this chapter (Figures 2.15 to 2.21) [20] and can be used with logs from other service companies. Complete chart books are available from most

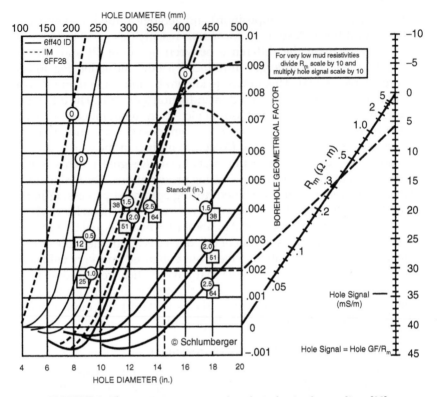

FIGURE 2.15   Borehole correction chart for induction log readings [20].

wireline service companies. Current chart books should be used and can be obtained by calling the appropriate service company or by asking the logging engineer. Correction and interpretation charts for order tools (normals and laterals) are no longer published by service companies, but can be found in a text by Hilchie [23] and in a recent publication [24]. All log readings should be corrected with these charts to obtain accurate R_t and R_{xo} readings.

The dual induction logs should be corrected for borehole, bed thickness, and invasion effects if three curves are present. The induction-SFL, induction-LL-8, and induction-electric log combinations require an R_{xo} curve to make corrections. If no R_{xo} device is presented, the deep induction curve is assumed to read true R_t. The specific charts included in this text are Figure 2.15 for induction log borehole corrections, Figures 2.16 and 2.17 for induction log bed-thickness corrections, and Figure 2.18a to 2.18c for invasion corrections (check log to see if an SFL or a LL-8 log was used with the induction curves and select the appropriate Tornado chart). Phasor induction logs only require invasion correction. This is accomplished with the appropriate Tornado Charts (Figures 2.18d to 2.18k) [20]. An invasion

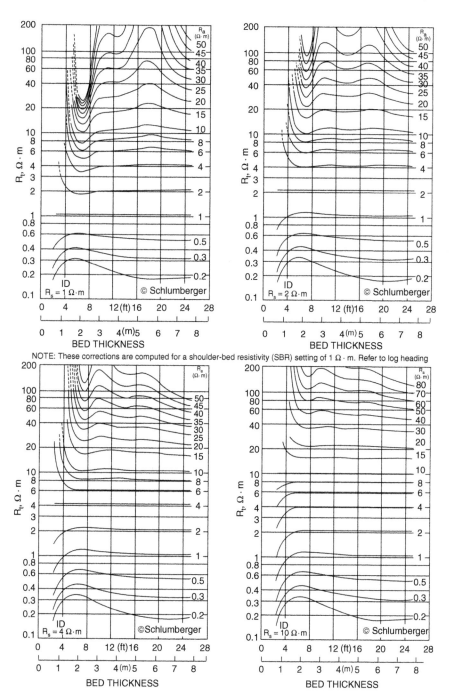

FIGURE 2.16 Bed thickness correction charts for the deep induction log [20].

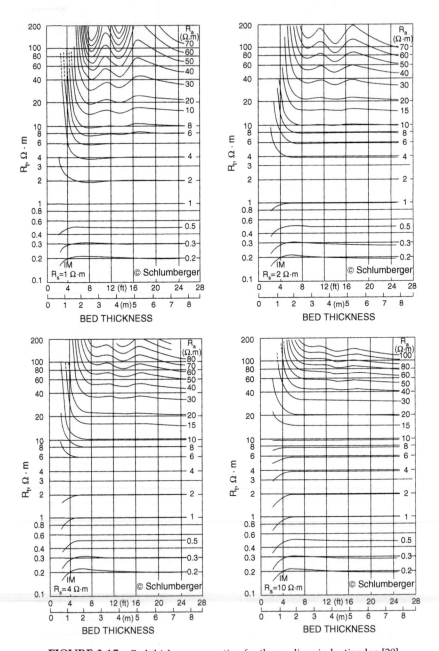

FIGURE 2.17   Bed thickness correction for the medium induction log [20].

(a)

Thick Beds, 8-in. (203-mm) Hole, Skin-Effect Corrected,
$R_{xo}/R_m \approx 100$, DIS-DB or Equivalent

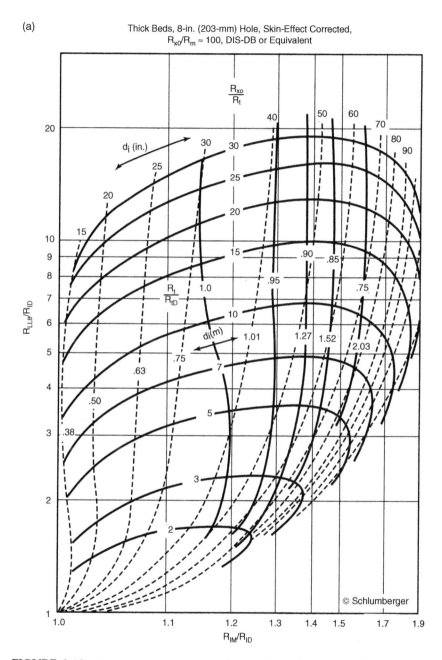

FIGURE 2.18a Invasion correction charts for the dual induction-laterolog 8 combination [20].

(b)   Thick Beds, 8-in. (203-mm) Hole, Skin-Effect Corrected,
                       $R_{xo}/R_m \approx 20$, DIS-DB or Equivalent

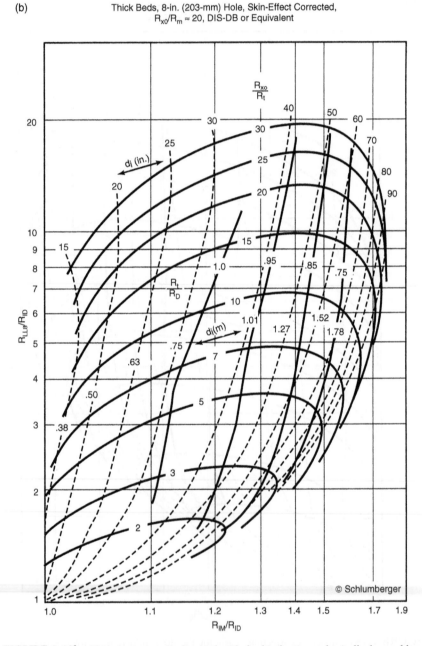

FIGURE 2.18b   This chart may also be used with dual induction-spherically focused log.

(c)

Thick Beds, 8-in. (203-mm) Hole, Skin-Effect Corrected,
$R_{xo}/R_m \approx 100$, DIS-EA or Equivalent

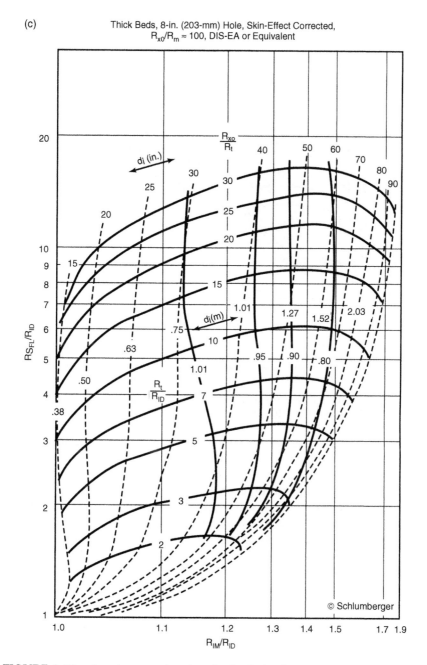

FIGURE 2.18c Invasion correction chart for the dual induction-spherically focused log combination [20].

(d)

Thick Beds, 8-in. (203-mm) Hole, Skin-Effect and Borehole Corrected,
$R_{xo}/R_m \approx 100$, DIT-E or Equivalent, Frequency = 20 kHz

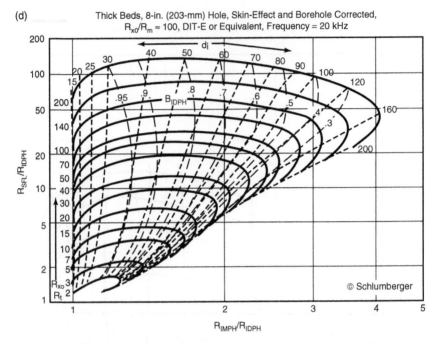

These Charts (Figures 2.18d and e) apply to the Phasor Induction tool when operated at a frequency of 20 kHz.
Similar charts (not presented here) are available for tool operation at 10 kHz and 40 kHz.
The 20 kHz charts do provide, however, reasonable approximations of R_{xo}/R_t and R_t/R_{IDPH} for tool operation at
10 kHz and 40 kHz when only moderately deep invasion exists (less than 100 inches)

FIGURE 2.18d Invasion correction chart for the phasor dual induction-spherically focused
log combination for Rxo/Rm = 100 [20].

correction Tornado Chart for phasor induction tools with other frequencies
are also available [24].

The dual laterolog-R_{xo} combination should be corrected for borehole
effects; bed thickness corrections are not normally made. Figure 2.19a is
used to make the borehole corrections of the deep and shallow laterologs,
respectively [20]. Figure 2.20 is used to correct the micro-SFL logs for mud
cake effects [20]. Invasion effects are corrected by using Figure 2.21 [20].

2.3.4.8 Interpretation

With Equation 2.20, water saturation can be found using corrected values
of R_t derived from the Tornado charts, R_w can be found from an SP log or
chemical analysis, and F_R from a porosity log and Table 1.7.

If a porosity log is not available, F_R can be found using Equation 1.45.
R_o is selected in a water-bearing zone from the deepest reading resistivity
curve. Care should be taken to select R_o in a bed that is thick, permeable,
and clean (shalefree). The zone should be as close to the zone of interest as
possible. R_o can sometimes be selected from below the water/hydrocarbon

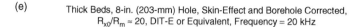

(e)

Thick Beds, 8-in. (203-mm) Hole, Skin-Effect and Borehole Corrected,
$R_{xo}/R_m \approx 20$, DIT-E or Equivalent, Frequency = 20 kHz

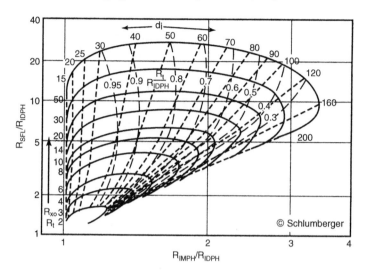

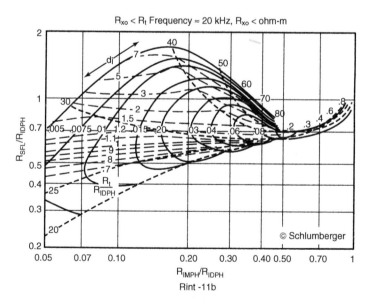

FIGURE 2.18e   Invasion correction chart for the phasor dual induction-spherically focused log combination for Rxo/Rm = 20 [20].

2. FORMATION EVALUATION

(f)

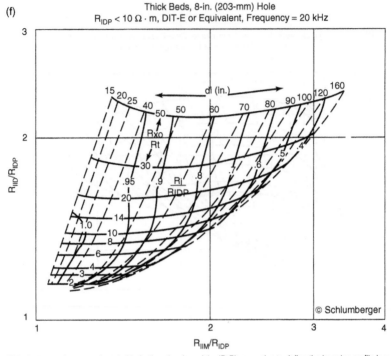

This chart uses the raw, unboosted induction signals and the ID-Phasor value to define the invasion profile in a rock drilled with oil-base mud. To use the chart, the ratio of the raw, unboosted medium induction signal (IIM) and the deep Phasor induction (IDP), is entered in abscissa. The ratio of the raw, unboosted deep induction signal (IID) and the deep Phasor induction (IDP) is entered in ordinate. Their intersection defines d_i, R_{xo}/R_t, and R_t.

EXAMPLE:
$R_{IDP} = 1.6 \,\Omega \cdot m$
$R_{IID} = 2.4 \,\Omega \cdot m$
$R_{IIM} = 2.4 \,\Omega \cdot m$
Giving, $R_{IID}/R_{IDP} = 2.4/1.6 = 1.5$
$R_{IIM}/R_{IDP} = 2.4/1.6 = 1.5$
Therefore, $d_i = 50$ in.
$R_{xo}/R_t = 15$
$R_t/R_{IDP} = 0.94$
$R_t = 0.94 \,(1.6)$
$= 1.5 \,\Omega \cdot m$

FIGURE 2.18f Invasion correction chart for the phasor dual induction log in oil-based mud [20].

contact within the zone of interest. Archie's equation (as described earlier) can then be used to calculate water saturation:

$$S_w = \left[\frac{F_R R_w}{R_t} \right]^{1/2}$$

If no porosity log or water zone is available, the ratio method can be used to find water saturation. The saturation of the flushed zone (S_{xo}) can

(g)

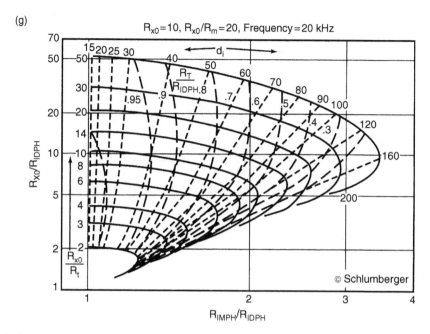

FIGURE 2.18g Invasion correction chart for the phasor dual induction-Rxo log combination for Rxo = 10, Rxo/Rm = 20 [20].

(h)

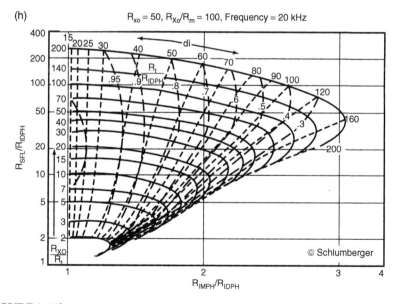

FIGURE 2.18h Invasion correction chart for the phasor dual induction-Rxo log combination for Rxo = 50, Rxo/Rm = 100 [20].

(i)

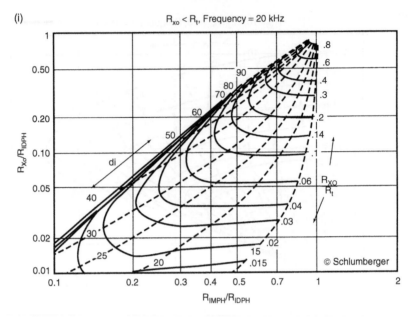

FIGURE 2.18i   Invasion correction chart for the phasor dual induction-Rxo log combination for Rxo < Rt [20].

(j)

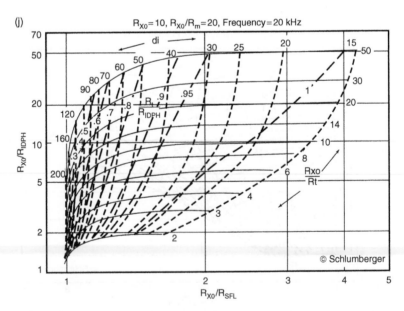

FIGURE 2.18j   Invasion correction chart for the phasor dual induction-spherically focused-Rxo log combination for Rxo = 10, Rxo/Rm = 20 [20].

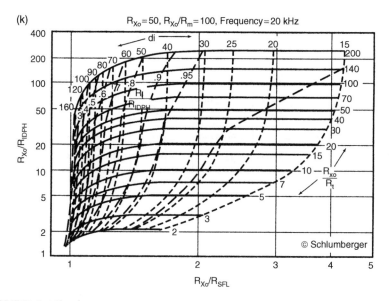

FIGURE 2.18k Invasion correction chart for the phasor dual induction-spherically focused-Rxo log combination for Rxo = 50, Rxo/Rm = 100 [20].

be found by:

$$S_{xo} = \left[\frac{FR_{mf}}{R_{xo}}\right]^{1/2} \tag{2.19}$$

When Equation 1.70 is divided by Equation 2.19, the result is a ratio of water saturations:

$$\frac{S_w^2}{S_{xo}^2} = \frac{R_{xo}R_w}{R_t R_{mf}} \tag{2.20}$$

Poupon, Loy, and Tixier [25] found that for "average" residual oil saturations:

$$S_{xo} = S_w^{1/5} \tag{2.21}$$

Substituting Equation 2.21 into Equation 2.20 and rearranging terms:

$$S_w = \left[\frac{R_{xo}/R_t}{R_{mf}/R_w}\right]^{5/8} \tag{2.22}$$

With this equation, water saturation can be found without knowing ϕ. Note, however, that this interpretation method is based on the assumption that $S_{xo} = S_w^{1/5}$. This relation is for "average" granular rocks and may vary considerably in other rock types. Figure 2.22 is a chart [20] that solves Equation 2.22.

(a)

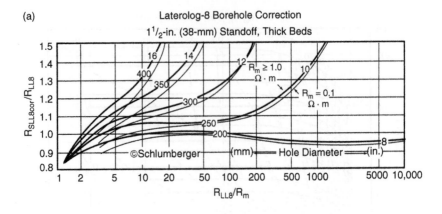

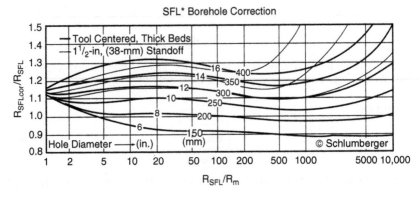

FIGURE 2.19a Borehole correction charts for the laterolog and spherically focused logs [20].

When two or more resistivity logs with different depths of investigation are combined, permeable zones can be identified. In a permeable zone, the area closest to the borehole will be flushed of its original fluids; mud filtrate fills the pores. If the mud filtrate has a different resistivity than the original formation fluids (connate water), the shallowest-reading resistivity tool will have a different value than the deepest-reading tool (Figure 2.23). Many times this difference is significant. The separation of the resistivity curves that result is diagnostic of permeable zones.

Care should be taken not to overlook zones in which curves do not separate. Curve separation may not occur if:

1. The mud filtrate and original formation fluids (i.e., connate water and hydrocarbons) have the same resistivity; both shallow and deep tools will read the same value. This is usually not a problem in oil or gas-saturated rocks.

(b)

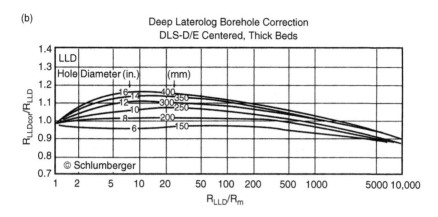

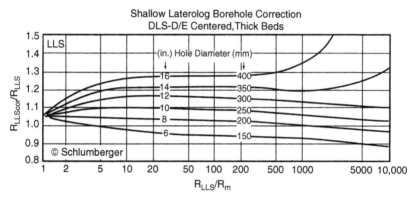

FIGURE 2.19b Borehole correction charts for the dual laterolog [20].

2. Invasion of mud filtrate is very deep, both shallow and deep tools may read invaded-zone resistivity. This occurs when a long period of time elapses between drilling and logging or in a mud system with uncontrolled water loss.

Microresistivity Tools   Microresistivity tools are used to measure the resistivity of the flushed zone. This measurement is necessary to calculate flushed zone saturation and correct deep-reading resistivity tools for invasion. Microresistivity tools are pad devices on hydraulically operated arms. The microlog and proximity log are the two main fresh-mud microresistivity tools, while the micro SFL and microlaterolog are the two main salt-mud microresistivity tools.

Figure 2.24 [21, 26, 52] shows the electrode pads and current paths for the Microlog (2.24a), Micro SFL. (2.24b), and proximity log (2.24c), and Microlaterolog. Figure 2.24d shows the Micro SFL. sonde.

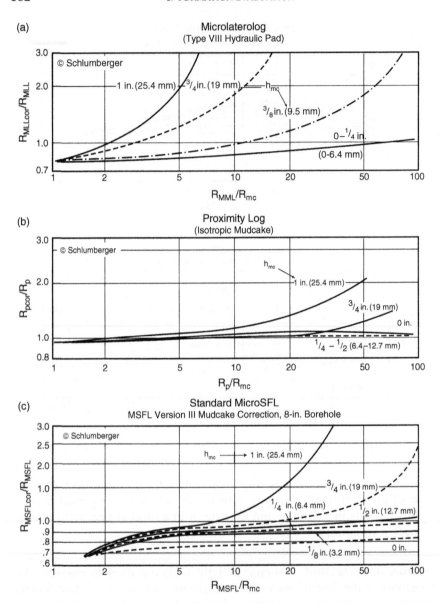

FIGURE 2.20 Mudcake correction for: (a) microlaterolog, (b) proximity log, and (c) micro SFL log.

2.3.4.9 Theory

The microlog makes two shallow nonfocused resistivity measurements, each at different depths. The two measurements are presented simultaneously on the log as the micronormal and microinverse curves. Positive

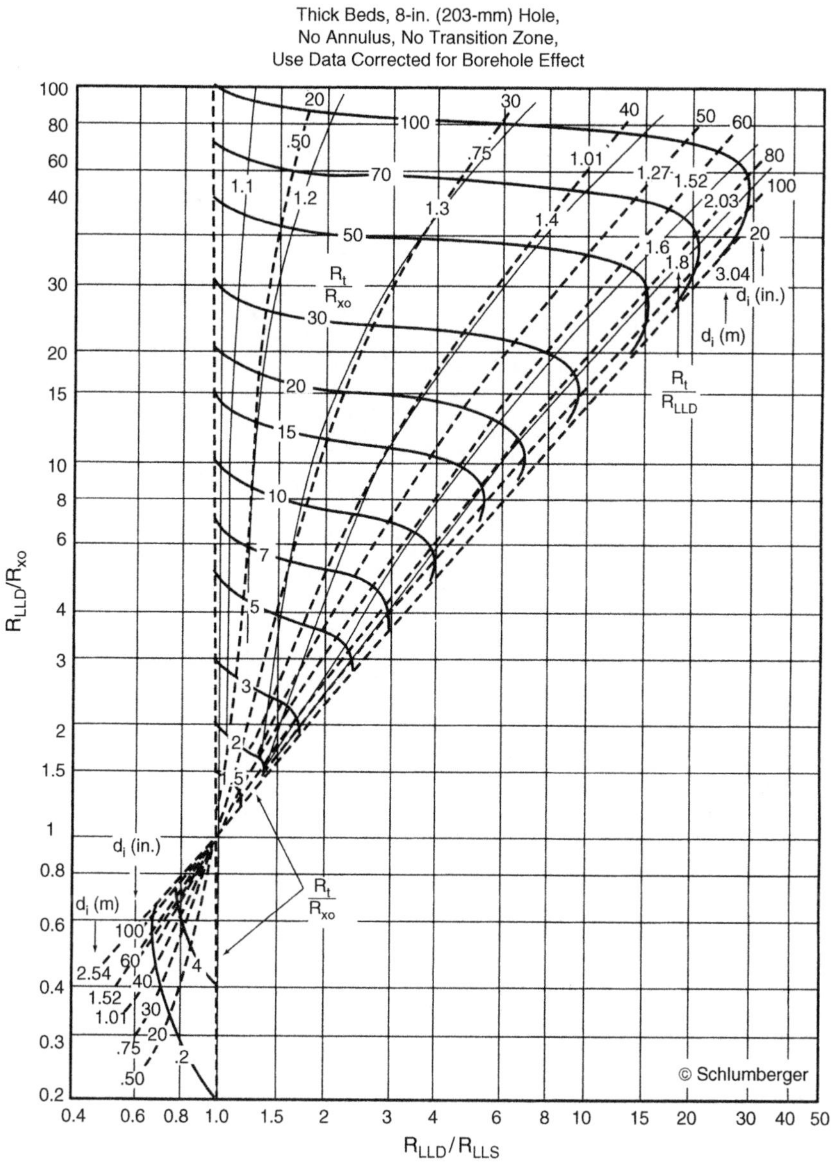
Thick Beds, 8-in. (203-mm) Hole,
No Annulus, No Transition Zone,
Use Data Corrected for Borehole Effect
R_t/R_{xo}
R_t/R_{LLD}
d_i (in.)
d_i (m)
R_{LLD}/R_{xo}
R_{LLD}/R_{LLS}
© Schlumberger

FIGURE 2.21 Invasion correction chart for the dual laterolog-R_{xo} log combination [20].

separation (micronormal reading higher than the microinverse) indicates permeability. R_{xo} values can be found by using Figure 2.25 [20]. To enter Figure 2.25, R_{mc} must be corrected to formation temperature, and mudcake thickness (h_{mc}) must be found. To find h_{mc} subtract the caliper reading (presented in track 1) from the borehole size and divide by two. In washed out or

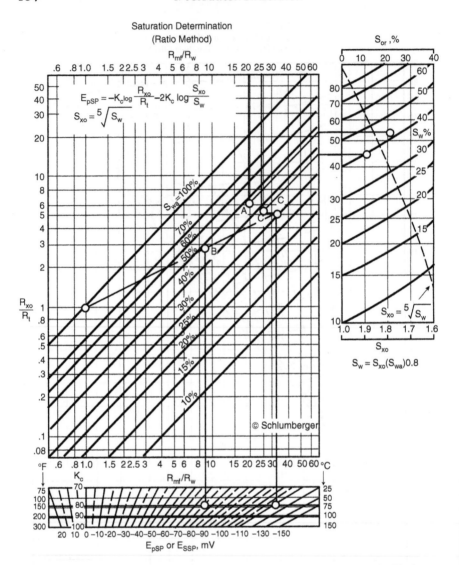

FIGURE 2.22   Chart for determining water saturation by the ratio method [20].

enlarged boreholes, h_{mc} must be estimated. The microlog gives a reasonable value of R_{xo} when mudcake thickness is known. The main disadvantage of this tool is that it cannot be combined with an R_t device, thus a separate logging run is required. The microlog is primarily a fresh mud device and does not work very well in salt-based muds [27]. Generally the mudcake is not thick enough in salt-based muds to give a positive separation opposite permeable zones. The backup arm on the microlog tool, which provides a caliper reading, is also equipped with either a proximity log

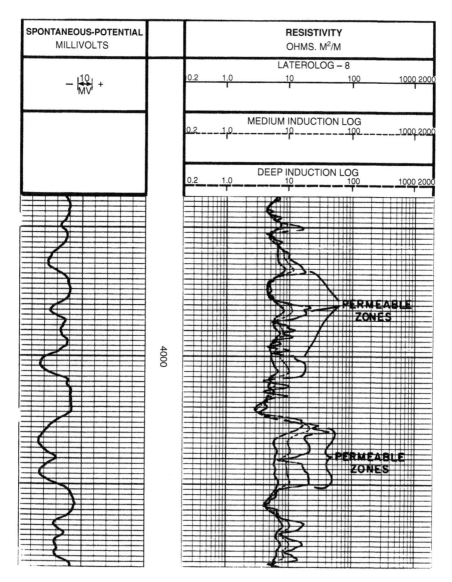

FIGURE 2.23 Example of resistivity log curve separation as an indication of permeability.

or a Microlaterolog. The proximity log is designed for fresh muds where thick mudcakes develop opposite permeable formations. There is essentially no correction necessary for mudcakes less than $\frac{3}{4}$ in. If invasion is shallow, the R_{xo} measurement may be affected by R_t because the proximity log reads deeper into the formation than the other microresistivity logs. When the microlog is run with a proximity log, it is presented in track 1

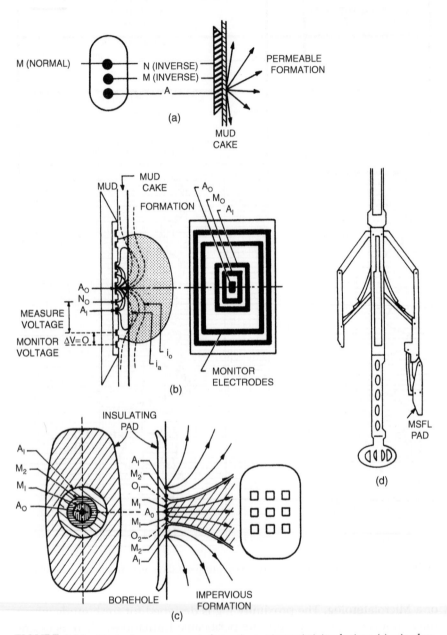

FIGURE 2.24   Electrode arrangements for various microresistivity devices: (a) microlog; (b) microspherically focused log; (c) microlaterolog; (d) location of the micro pad on the dual laterolog sonde [21, 26, 52].

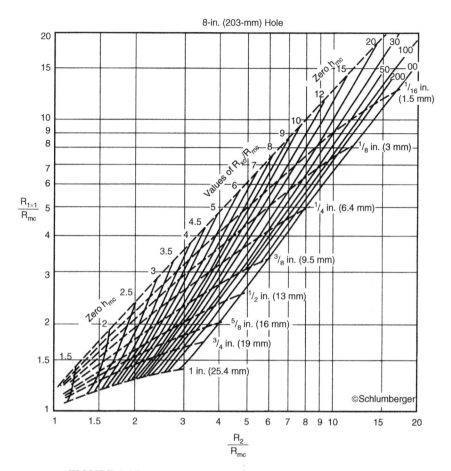

8-in. (203-mm) Hole

FIGURE 2.25 Chart for finding Rxo from microlog readings [20].

with a microcaliper. The proximity log is presented in tracks 2 and 3 on a logarithmic scale (Figure 2.26).

Since its introduction, the microlog (Schlumberger) has become the standard tool for recognizing permeable zones. The theory behind it is similar to using multiple resistivity devices. The tool consists of three electrode buttons on a rubber pad which is pressed against the borehole wall.

In a permeable zone, mud filtrate will enter the formation leaving the clay particles behind on the borehole wall. These clay particles may form a mudcake up to an inch thick. The resistivity of the mudcake is less than the resistivity of the formation saturated with mud filtrate. Two resistivity readings, the microinverse and the micronormal, are taken simultaneously. The microinverse has a depth of investigation of only an inch; therefore, it reads mostly mudcake (if present). The micronormal has a depth of investigation

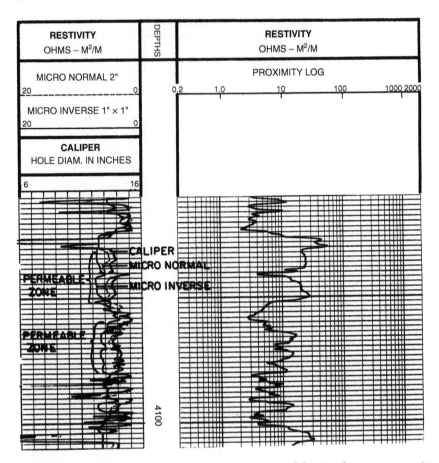

FIGURE 2.26 Example log showing positive separation of the microlog curves opposite a permeable formation at 4,030 to 4,050 feet.

of 3 to 4 in. and is influenced primarily by fluids in the flushed zone. The difference in resistivity shows up on the log as a separation of the curves with the micronormal reading higher than the microinverse. This is referred to as "positive separation." In impermeable formations, both readings are very high and erratic, and negative separation may occur (micronormal less than microinverse). Shales commonly show negative separation with low resistivities (Figure 2.27).

In salt muds, the microlaterolog and micro spherically focused log (MSFL) are used for R_{xo} readings. The microlaterolog is a focused tool with a shallower depth of investigation than the proximity log. For this reason, the microlaterolog is very strongly affected by mudcakes thicker than $\frac{3}{8}$ in. It is presented in tracks 2 and 3 like the proximity log. The MSFL is the most common R_{xo} tool for salt muds. It is a focused resistivity device that

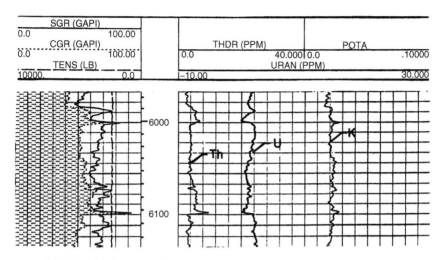

FIGURE 2.27   Example of a natural gamma spectroscopy log presentation.

can be combined with the dual laterolog, thus providing three simultaneous resistivity readings. Although the depth of investigation is only a few inches, the tool can tolerate reasonably thick mudcakes $\frac{3}{4}$ in. The tool is also available in a slim-hole version. The only disadvantage to this device is that the pad can be easily damaged in rough boreholes.

2.3.4.10   Interpretation

The saturation of the flushed zone can be found from Equation 2.20. R_{mf} must be at formation temperature. Moveable hydrocarbons can be found by comparing S_{xo} and S_w. If $S_w/S_{xo} < 0.7$ then the hydrocarbons in the formation are moveable (this is also related to fluid permeability). If $S_w/S_{xo} > 0.7$, either there are no hydrocarbons or the hydrocarbons present are not moveable.

Gamma Ray Logs   The gamma ray log came into commercial use in the late 1940s. It was designed to replace the SP in salt muds and in air-filled holes where the SP does not work. The gamma ray tool measures the amount of naturally occurring radioactivity in the formation. In general, shales tend to have high radioactivity while sandstone, limestone, dolomite, salt, and anhydrite have low radioactivity. There are exceptions. Recently, tools have been designed to separate gamma rays into their respective elemental sources, potassium (K), thorium (Th), and uranium (U).

2.3.4.11   Theory

Gamma rays are high-energy electromagnetic waves produced by the decay of radioactive isotopes such as K40, Th, and U. The rays pass from the formation and enter the borehole. A gamma ray detector (either scintillation detector or Geiger-Muller tube) registers incoming gamma rays as an electronic pulse. The pulses are sent to the uphole computer where they are counted and timed. The log, presented in track 1 in Figure 2.13, is in API units.

As previously mentioned, there are new gamma ray tools available that determine which elements are responsible for the radioactivity. The incoming gamma rays are separated by energy levels using special energy-sensitive detectors. The data are collected by the computer and analyzed statistically. The log presents total (combined) gamma ray in track 1 and potassium (in %), and uranium and thorium (in ppm) in tracks 2 and 3 (Figure 2.27). Combinations of two components are commonly presented in track 1. The depth of investigation of the natural gamma tools is 2–10 in. depending on mud weight, formation density, hole size, and gamma ray energies.

2.3.4.12 Interpretation

The interpretation of a total gamma ray curve is based on the assumption that shales have abundant potassium-40 in their composition. The open lattice structure and weak bonds in clays encourage incorporation of impurities. The most common of those impurities are heavy elements such as uranium and thorium. Thus, shales typically have high radioactivity. Sandstones (quartzose), carbonates and evaporites have strong bonds and generally do not allow impurities. Limestones undergo rearrangement of crystal structure and addition of magnesium to become dolomites. Impurities like uranium (which is very soluble) may enter the crystal lattice during recrystallization. Feldspathic sandstones contain an abundance of potassium-40 and therefore show higher radioactivity than quartzose sandstone. Some evaporite minerals (Such as KCl) contain high amounts of potassium-40 and may appear as shales on the log. Serra et al. [28] provide an excellent discussion of interpretation of the natural or spectral gamma-ray tool.

Sonic (Acoustic) Log The sonic (acoustic or velocity) total measures the time it takes for a compressional wave to travel through one vertical foot of formation. It can be used to determine porosity (if the lithology is known) and to determine seismic velocities for geophysical surveys when combined with a density log. The sonic log also has numerous cased hole applications.

2.3.4.13 Theory

A 20 kHz sound wave is produced by the tool and travels through the mud into the formation. The wave travels vertically through the formation. The first arrival of the compressional wave is picked up by a received about one foot away from the transmitter. The wave continues through the formation and is picked up by the far receiver (normally 2-ft below the near receiver). The time difference between the near and far receivers is used to determine formation travel time (Δt). Fractures, vugs, unconsolidated formations, gas-cut mud, lost circulation materials, and rough boreholes can cause sharp increases in Δt, called cycle skips.

2.3.4.14 Interpretation

Table 2.7 shows the velocity and travel time for several commonly encountered oilfield materials. The t_{ma} value in the fourth column is at 0% porosity. Porosity increases travel time. Wyllie and coworkers [208] developed an equation that relates since travel time to porosity:

$$\phi = \frac{\Delta t_{log} - \Delta t_{ma}}{\Delta t_f - \Delta t_{ma}} \tag{2.23}$$

where $\Delta t_{log} = \Delta t$ value read from log, $\mu\,sec/ft$
$\Delta t_{ma} = $ matrix velocity at 0% porosity, $\mu\,sec/ft$
$\Delta t_f = 189\text{--}190\ \mu\,sec/ft$ (or by experiment)

The Wyllie equation works well in consolidated formations with regular intergranular porosity ranging from 5%–20% [30]. If the sand is not consolidated or compacted, the travel time will be too long, and a compaction correction factor (C_p) must be introduced [29]. The reciprocal of C_p is multiplied by the porosity from the Wyllie equation:

$$\phi = \left(\frac{\Delta t_{log} - \Delta t_{ma}}{t_f - t_{ma}} \right) \left(\frac{1}{C_p} \right) \tag{2.24}$$

The compaction correction factor (C_p) can be found by dividing the sonic porosity by the true (known) porosity. It can also be found by dividing the travel time in an adjacent shale by 100:

$$C_p = \frac{\Delta t_{shale}}{100} C \tag{2.25}$$

TABLE 2.7   Matrix Travel Times

Material	Velocity range ft/sec	Δt range $\mu\,sec/ft$	Δt commonly used $\mu\,sec/ft$	Δt at 10% porosity $\mu\,sec/ft$
Sandstone	18,000–19,500	51.0–55.5	55.0 or 51.0	69.0 or 65.0
Limestone	21,000–23,000	43.5–47.6	47.5	61.8
Dolomite	23,000–24,000	41.0–43.5	43.5	58.0
Salt	15,000	66.7	66.7	—
Anhydrite	20,000	50.0	50.0	—
Shale	7,000–17,000	58.0–142	—	—
Water	5,300	176–200	189	—
Steel casing	17,500	57.0	57.0	—

From References 21 and 36.

where C is a correction factor, usually 1.0 [21]. In uncompared sands, porosities may be too high even after correction if the pores are filled with oil or gas. Hilchie [21] suggests that if pores are oil-filled, multiply the corrected porosity by 0.9; if gas-filled, use 0.7 to find corrected porosity.

Raymer, Hunt, and Gardner [31] presented an improved travel-time to porosity transform that has been adopted by some logging companies. It is based on field observations of porosity versus travel time:

$$\Delta t_{log} = (1 - \phi)^2 \Delta t_{ma} + \phi \Delta t_f \tag{2.26}$$

This relationship is valid up to 37% porosity.

The heavy set of lines in Figure 2.28 [20] was derived using the Raymer et al. [31] transform, and the lighter set represents the Wyllie relationship.

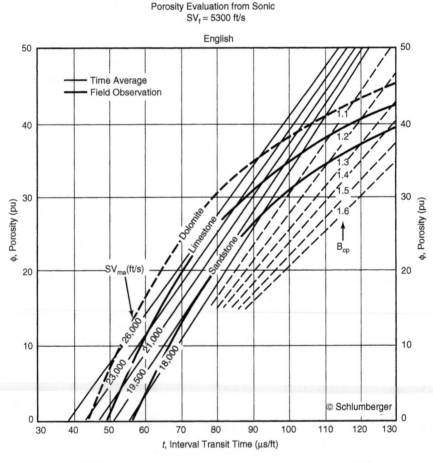

FIGURE 2.28   Chart for evaluating porosity with a sonic log [20].

The sonic porosity derived from the Wyllie equation (Equation 2.23) does not include secondary porosity (vugs and fractures), so it must be cautiously applied in carbonate rocks.

Density Log   The formation density tool measures the bulk electron density of the formation and relates it to porosity. It is a pad device with a caliper arm. The tool is usually run in combination with a neutron log, but it can be run alone.

2.3.4.15 Theory

The density tool emits medium energy gamma rays from a radioactive chemical source (usually Cs-137). The gamma ray penetrate the formation and collide with electron clouds in the minerals in the rock. With each collision the gamma ray loses some energy until it reaches a lower energy state. This phenomenon is called "Compton scattering." Some gamma rays are absorbed, and a high-energy electron is emitted from the atom. This phenomenon is called the "photoelectric absorption" effect, and is a function of the average atomic weight of each element. Both the Compton-scattered gamma ray and the photo-electrically produced electron return to the borehole where they are detected by scintillation tubes on the density tool. The main result is that a porous formation will have many returning gamma rays while a nonporous formation will have few returning gamma rays. Each tool has two detectors; one is near the source (short-spacing detector) and another is 1–1.5 ft (35–40 cm) away from the near detector (long-spacing detector).

The long-spacing detector provides the basic value of bulk electron density. The short-spacing detector is used to make a mudcake correction. This correction, made automatically by computer by most service companies, is based on the "spine and ribs" plot (Figure 2.29) [5]. The "spine" is the heavy, nearly vertical line from 1.9 to 2.9 g/cc. The ribs are the lighter curved lines trending left to right. The experimental data for constructing the "ribs" are shown in the corners of the plot. Long-spacing count-rates are on the abscissa and short-spacing count-rates are on the ordinate axis. The computer receives data from the sonde and plots it on the chart. If the point falls off the "spine," it is brought back along one of the ribs. Bringing the point back along the "rib" will change the intersection point on the "spine." The correction that is produced is called $\Delta\rho$ and may be either positive or negative depending on the mud properties. Negative $\Delta\rho$ values occur in heavy (barite or iron), weighted muds. Positive values occur in light muds and when the density pad is not flush against the borehole wall as occurs in rough or "rugose" boreholes. The $\Delta\rho$ curve is useful for evaluating the quality of the ρ_b reading. Excursions from 0 that are more than ± 0.20 gm/cc on the $\Delta\rho$ curve indicate a poor quality reading.

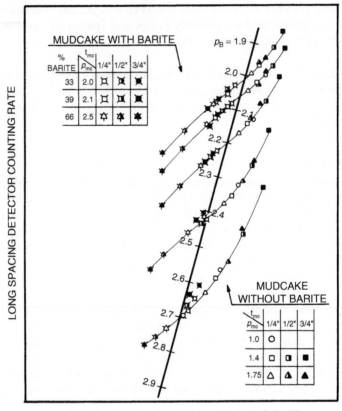

FIGURE 2.29 Spine and ribs plot used to correct bulk density readings for mudcake effects [52].

2.3.4.16 Interpretation

An equation similar to the Wyllie equation is used to calculate porosity values from bulk density.

$$\phi = \frac{\rho_{ma} - \rho_b}{\rho_{ma} - \rho_f} \tag{2.27}$$

where ρ_{ma} = bulk density of matrix at 0% porosity, g/cc
  ρ_b = bulk density from log, g/cc
  ρ_f = bulk density of fluid, g/cc

Table 2.8 lists commonly used values for ρ_{ma} and ρ_f, and, along with Figure 2.30 shows how ρ_f changes with temperature and pressure. As with the sonic tool, an incorrect choice of matrix composition may give negative porosity values.

TABLE 2.8   Bulk Densities Commonly Used for Evaluating Porosity With a Density Log*

Material	ρ_{bulk}	ρ_{log} at 10% porosity (fresh water)
Unconsolidated sand	2.65 g/cc	2.48 g/cc
Silica cemented sand	2.65 g/cc	2.48 g/cc
Calcite cemented sand	2.68 g/cc	2.51 g/cc
Limestone	2.71 g/cc	2.54 g/cc
Dolomite	2.83–2.87 g/cc	2.64–2.68 g/cc
Salt	2.03 g/cc	—
Anhydrite	2.98 g/cc	—
Fresh water	1.0 g/cc	—
Salt water	1.1–1.2 g/cc	—

Fluid Densities for Water (Based on Salinity)**	
Salinity, ppm Nacl	ρ_f, g/cc
0–50,000	1.0
50,000–100,000	1.03
100,000–150,000	1.07
150,000–200,000	1.11
200,000–250,000	1.15
250,000–300,00	1.19

*From Reference 36.
**From Reference 21.

If a zone is hydrocarbon saturated but not invaded by mud filtrates, the low density of the hydrocarbons will increase the porosity reading to a value that is too high. In this case, Hilchie [21] suggests using the following equation:

$$\phi = \frac{0.9 \left(\frac{R_w}{R_t}\right)^{1/2} (\rho_w - \rho_h) + \rho_{ma} - \rho_b}{\rho_{ma} - \rho_h} \qquad (2.28)$$

where ρ_w = density of formation water, g/cc (estimated from Figure 2.30)
ρ_h = density of hydrocarbons, g/cc (from Figure 2.31) [20]

Neutron Log   Neutron tools measure the amount of hydrogen in the formation and relate it to porosity. High hydrogen content indicates water (H_2O) or liquid hydrocarbons (C_xH_z) in the pore space. Except for shale, sedimentary rocks do not contain hydrogen in their compositions.

2.3.4.17 Theory
Neutrons are electrically neutral particles with mass approximately equal to that of a hydrogen atom. High-energy neutrons are emitted from a

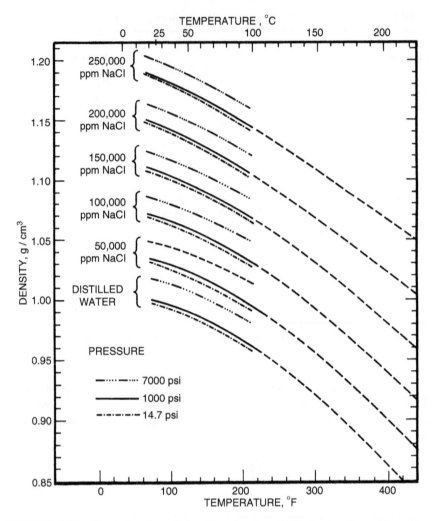

FIGURE 2.30 Chart showing relationship of water salinity to density and temperature [58].

chemical source (usually AmBe or PuBe). The neutrons collide with nuceli of the formation minerals in elastic-type collisions. Neutrons will lose the most energy when they hit something with equal mass, such as a hydrogen atom. A few microseconds after being released, the neutrons have lost significant energy and enter the thermal state. When in the thermal state, neutrons are captured by the nuclei of other atoms (Cl, H, B). The atom which captures the neutron becomes very excited and emits a gamma ray. The detectors on the tool may detect epithermal neutrons, thermal neutrons

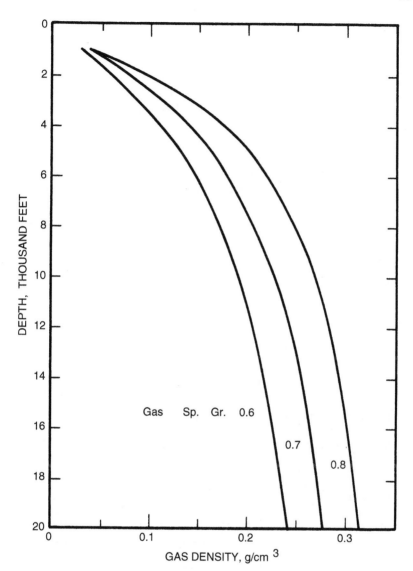

FIGURE 2.31 Chart for estimating density of gases from reservoir depth and gas specific gravity [21].

or high-energy gamma rays of capture. Compensated neutron tools (CNL) detect thermal neutrons and use a ratio of near-to-far detector counts to determine porosity. Sidewall neutron tools (SNP) detect epithermal neutrons and have less matrix effect (though they are affected by rough boreholes more than the CNL).

2.3.4.18 *Interpretation*

Neutron tools are seldom run alone. They are usually combined with a density-porosity tool. Older neutron logs are not presented as porosity but as count rates. Some logs do not specify a scale (Figure 2.32), but only which direction the count rate (or radiation) increases. An increase in radiation indicates lower porosity (less hydrogen). Newer logs present porosity (for a particular matrix, limestone, sandstone or dolomite) directly on the log. Most neutron logs are run on limestone matrix. Figure 2.33 corrects the porosity for matrix effect if the log is run on limestone matrix [20]. Neutron logs exhibit "excavation effect" in gas-filled formations. The apparent decrease in porosity is due to the spreading out of hydrogen in

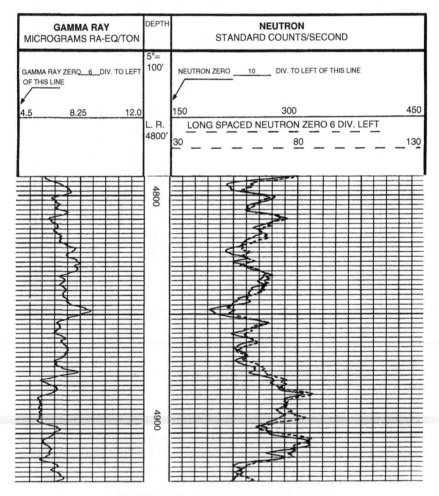

FIGURE 2.32 Old neutron log presentation.

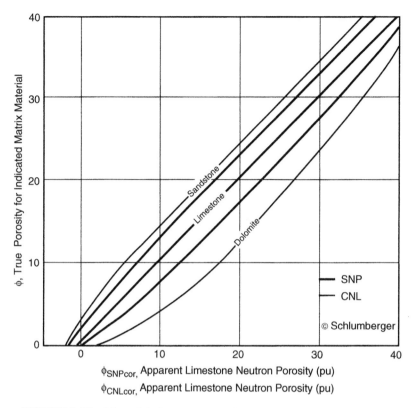

FIGURE 2.33 Matrix lithology correction chart for neutron porosity logs [20].

gas molecules; gases have less hydrogen *per unit volume* than liquids. Thus the neutron tool sees less hydrogen and assumes less porosity. The magnitude of the effect depends on gas saturation, gas density, and pressure. Care should be taken in using correction charts for neutron tools; each service company has a slightly different design, and the correct chart for the particular tool and service company should be used.

Multiple-Porosity Log Interpretation As mentioned earlier, the neutron, density, and sonic tools are lithology dependent. If the matrix is incorrectly selected, porosity may be off as much as 10 porosity units. If two lithology-dependent logs are run simultaneously on the same matrix and presented together, lithology and porosity can be determined. The most common and useful of combinations is the neutron and density. Figures 2.34 through 2.38 are crossplots of neutron, density, and sonic porosity. The charts are entered with the appropriate values on the ordinate and abscissa. The point defines porosity and gives an indication of matrix. If a point falls between two matrix lines, it is a combination of the two

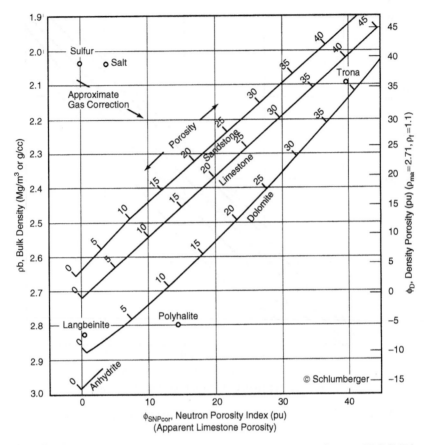

FIGURE 2.34   Chart for finding porosity and matrix composition from an FDC-SNP log combination in fresh water [20].

minerals or the neutron-density crossplots. Gas moves the points up and to the left. To correct for the gas effect, move parallel to the gas correction arrow to the assumed lithology. Note that a gassy limestone may look like a sandstone. Shales tend to bring points down and to the right depending on the shale composition. Typically, shaly sandstone will look like a limestone. The sonic-density crossplot is not very helpful in determining porosity or lithology but is extremely useful for determining evaporite mineralogy.

In some areas, neutron and density tools are run on to sandstone porosity and therefore cannot be entered in the charts directly. To use the neutron-density crossplots when the matrix is not limestone, another method must be applied. Remember that the vertical lines are constant neutron-porosity and the horizontal lines are constant density-porosity. Instead of entering the bottom or sides of the chart, select the appropriate lithology line in the

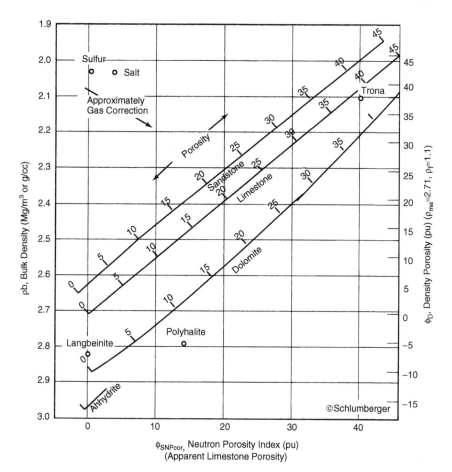

FIGURE 2.35 Chart for finding porosity and matrix composition from an FDC-SNP log combination in salt water [20].

interior of the chart. Draw a horizontal line through the density-porosity and a vertical line through the neutron-porosity. Lithology and porosity are determined at the intersection of these two lines.

Another device that provides good lithology and matrix control is the Lithodensity tool (LDT). It combines a density tool with improved detectors and a P_{ef} curve (photoelectric effect). Combining the P_b, and P_{ef} curve values, an accurate 3- or 4-mineral composition can be determined from the charts in Figures 2.39 to 2.42. This also provides an excellent way to confirm neutron-density cross-plot interpretations.

Nuclear Magnetic Resonance (NMR) This log examines the nucleus of certain atoms in the formation. Of particular interest are hydrogen nuclei (protons) since these particles behave like magnets rotating around each

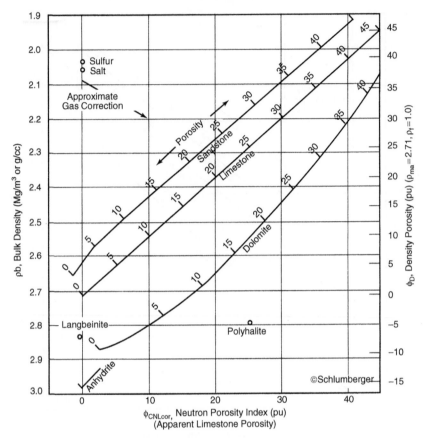

FIGURE 2.36 Chart for finding porosity and matrix composition from a FDC-CNL log combination in fresh water [20].

other [23]. Hydrogen is examined because it occurs in both water and hydrocarbons.

The log measures fluid by applying a magnetic field, greater in intensity than the earth field, to the formation. Hydrogen protons align themselves with the induced field and when the field is suddenly removed, the protons precess about the earth's magnetic field much like a gyroscope. The nucleus of hydrogen has a characteristic precession rate called the Larmor frequency ($\sim$ 2,100 Hz), and can be identified by sensors on the tool [24]. The nuclei contributing to the total signal occur in the free fluid in the pores; fluid adsorbed on the grains makes no contribution. The signals are then processed in a computer and printed out onto a log.

Normally, proton precession decays along a time constant, T_2. This is a result of each proton precession falling out of phase with other protons due to differences in the local magnetic fields. Moreover, each proton precesses

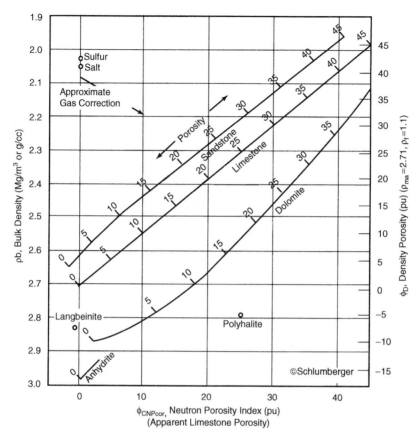

FIGURE 2.37 Chart for finding porosity and matrix composition from a FDC-CNL log combination in salt water [20].

at a slightly different frequency, depending on the kind of fluid it occurs in. This disharmonic relationship makes it possible to differentiate between free water and free oil in a reservoir [34].

Three log modes can be presented:

1. Normal mode—consists of the free fluid index (FFI) in percent obtained from a polarization time of two seconds, the Larmor proton frequency (LFRE), the decay-time constant (or longitudinal relaxation time) of the signal (T_2) and a signal-to-noise ratio (STNR).

2. Continuous mode—gives three free fluid index (FFI) readings taken at polarization times of 100, 200, and 400 ms, respectively, two longitudinal relaxation times (T_1 and T_2), and a signal-to-noise ratio (STNR).

3. Stationary display mode—a signal-stacking mode where eight signals are stacked from each of six polarization times to obtain precise T_1, T_2, and FFI values.

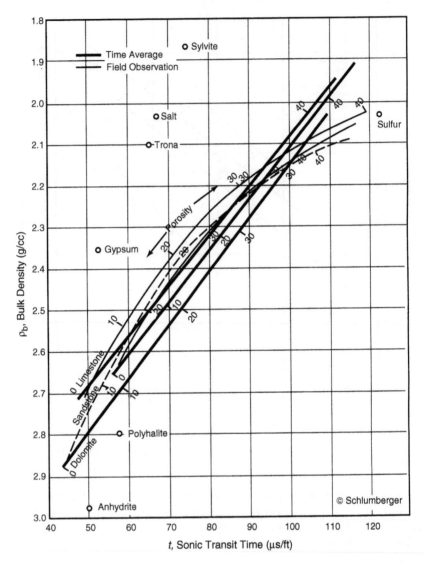

FIGURE 2.38 Lithology estimation chart for the FDC-sonic log combination in fresh water [20].

FFI readings yield porosity that is filled with moveable fluid and is related to irreducible water saturation (S_{iw}). Addition of paramagnetic ions to the mud filtrate will disrupt the water portion of the signal and residual oil saturation (S_{or}) can be determined.

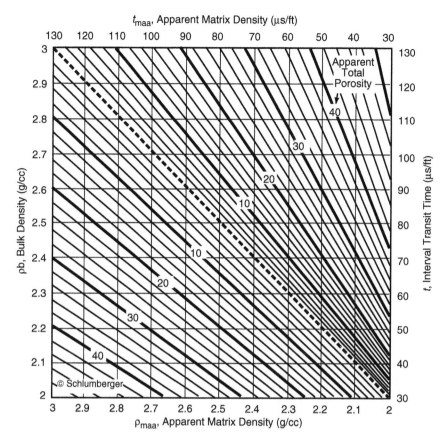

FIGURE 2.39 Chart for finding apparent matrix density or apparent matrix transit time from bulk density or interval transit time and apparent total porosity [20].

Desbrandes [33] summarized the following uses for the NML:

1. Measuring free fluid volume in the pores (ϕ_F).
2. Evaluating permeability by comparing ϕ_F with ϕ_T (total porosity from a neutron-density log combination).
3. Locating intervals at irreducible water saturation by comparing ϕ_F with ϕ_T and R_t, determined with other logs.
4. Determining residual oil saturation by adding paramagnetic ions to the mud filtrate to cancel the water signal and leave the oil signal.

Dielectric Measurement Tools   Dielectric measurement tools examine the formation with high frequency electromagnetic waves (microwaves)

2. FORMATION EVALUATION

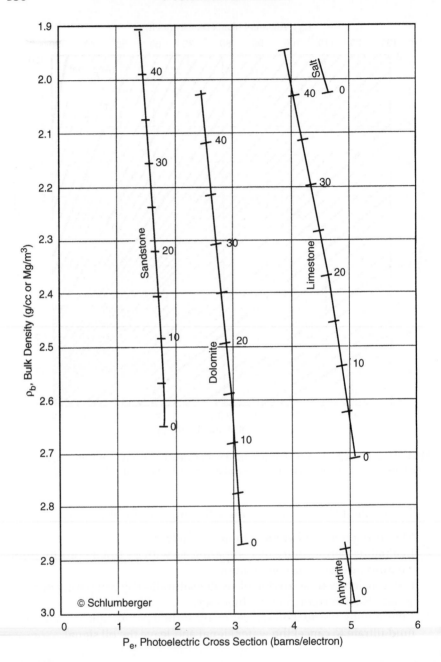

FIGURE 2.40 Chart for finding porosity and matrix composition from a lithodensity log in fresh water [20].

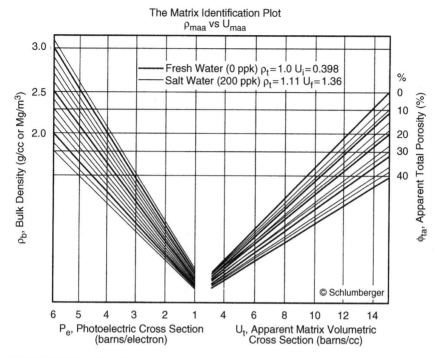

FIGURE 2.41 Chart for determining apparent matrix volumetric cross section from bulk density and photoelectric cross section [20].

rather than high-frequency sound waves (as in the sonic or acoustic logging tools). The way the electromagnetic wave passes through a given formation depends on the dielectric constants (ϵ) of the minerals and fluids contained in the rock.

Two types of tools are available [33]:

1. VHF sondes that have frequencies of 20–47 MHz (found in the deep propagation tool [Schlumberger], and dielectric tools [Dresser-Atlas and Gearhart-Owen], and
2. UHF sondes that have a frequency of 1.1 GHz (found in the electromagnetic propagation tool [Schlumberger]).

The only tool that is currently available is Schlumberger's electromagnetic propagation tool (EPT); the others are still experimental [33].

2.3.4.19 Theory

The EPT is a sidewall tool that measures the dielectric properties of a formation by passing spherically propagated microwaves into the rock. The tool consists of 2 transmitters (T_1, and T_2) and 2 receivers (R_1, and R_2) mounted in an antenna pad assembly. Its basic configuration is that of a

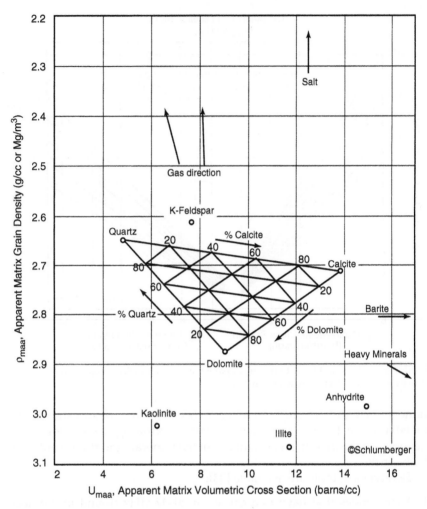

FIGURE 2.42 Matrix indentification crossplot chart for finding matrix composition from apparent matrix density and apparent matrix volumetric cross section [20].

borehole compensation array (much like the borehole compensated sonic (BHC) log). The transmitter fires a 1.1 GHz electromagnetic wave into the rock around the well-bore. As the wave passes through the rock and fluid there, it is attenuated, and its propagation velocity is reduced. The wave then refracts to the borehole where it is sequentially detected by the two receivers. How much the wave is attenuated is a function of the dielectric permittivity of the formation. Rocks and oil have similar permittivities while water has a very different permittivity. Therefore, the wave responds to the water-filled porosity in the formation, and the response is a function of formation temperature.

Since the wave is attenuated by water (and is not too bothered by oil), the log response indicates either R_{xo} (in water-based mud systems) or bulk volume water (in oil-based mud systems).

In order to provide usable values, the velocity of the returning wave is measured and compared to the wave-propagation velocity in free space. The propagation velocity of the formation is then converted into propagation time (T_{pl}). A typical log presentation includes a T_{pl}, curve (in nanoseconds/meter), an attenuation curve (EATT) in decibels/meter, and a small-arm caliper curve (which measures borehole rugosity) recorded in tracks 2 and 3. Figure 2.43 shows the basic antenna configuration [35]. Figure 2.44 is an example of an EPT log presentation.

The depth of investigation of the tool varies between one and three inches and depends on formation conductivity; high conductivity reduces depth of investigation.

The tool is affected primarily by hole roughness (rugosity) and mud cakes $> \frac{3}{8}$ in. thick. These effects reduce depth of investigation and in extreme situations (i.e., very rough holes and/or very thick mud cakes) keep the tool from reading the formation at all.

2.3.4.20 Interpretation

The most common way to interpret EPT logs is called the T_{po} *method* [35]. T_{po} in a clean formation is given by:

$$T_{po} = (\phi S_{xo} T_{pfo}) + \phi(1 - S_{xo})T_{phyd} + (1 - \phi)T_{pma} \qquad (2.29)$$

Rearranging terms and solving for S_{xo}:

$$S_{xo} = \frac{(T_{po} - T_{pma}) + \phi(T_{pma} - T_{phyd})}{\phi(T_{pfo} - T_{phyd})} \qquad (2.30)$$

where $T_{po} = T_{pl}$ corrected for conductivity losses, nanosecs/m
$\qquad T_{pma} =$ the matrix propagation time, nanosecs/m
$\qquad T_{phyd} =$ the hydrocarbon propagation time, nanosecs/m
$\qquad T_{pfo} =$ the fluid propagation time, nanosecs/m

T_{po} can be calculated by

$$T_{po} = T_{pl}^2 - \left(\frac{A_c}{60.03}\right)^2 \qquad (2.31)$$

where

$$A_c = A_{log} - 50 \qquad (2.32)$$

($A_{log} =$ EATT curve reading in dB/m)

Antenna configuration

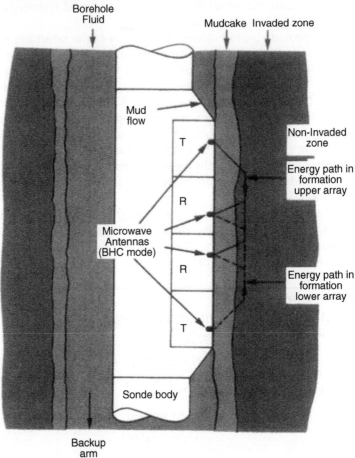

FIGURE 2.43 Diagram showing antenna-transmitter configuration for an EPT sonde and microwave ray paths through the mudcake and formation adjacent to the borehole [35].

T_{pfo} is a function of formation temperature (T) and can be found from:

$$T_{pfo} = \frac{20(710 - T/3)}{(440 - T/3)} \tag{2.33}$$

T_{pma} is taken from Table 2.9 and ϕ is taken from a neutron-density log(ϕ_{ND}). Equation 2.30 can be rearranged to find ϕ for a quick-look comparison with other porosity devices (specifically the neutron-density log). By assuming $T_{phyd} \approx T_{pfo}$, T_{phyd} can be eliminated and

$$S_{xo} = \frac{1}{\phi} \frac{(T_{po} - T_{pma})}{(T_{pfo} - T_{pma})} \tag{2.34}$$

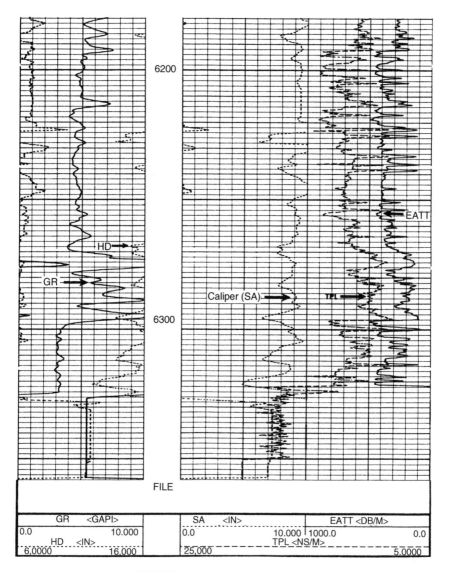

FIGURE 2.44 EPT log presentation.

Since the EPT log measures water-filled porosity, by definition

$$\phi = \phi_{EPT} \frac{(T_{po} - T_{pma})}{(T_{pfo} - T_{pma})}$$ (2.35)

in a water zone ($R_t = R_o$; $S_w = 100\%$)

TABLE 2.9 Matrix Propagation Times with the Electromagnetic Propagation Tool for Various Minerals

Mineral	T_{pma}, nanosec/m
Sandstone	7.2
Dolomite	8.7
Limestone	9.1
Anhydrite	8.4
Dry colloids	8.0
K-feldspar	7.0–8.2
Muscovite	8.3–9.4
Biotite	7.3–8.2
Talc	7.1–8.2
Halite	7.9–8.4
Siderite	8.8–9.1
Gypsum	6–8
Sylvite (KCl)	7.2–7.3
Limonite	10.5–11.0
Apatite	9.1–10.8
Sphalerite	9.3–9.6
Rutile	31.8–43.5
Petroleum	4.7–5.8
Shale	variable
Fresh water @ 250°C	29.0

From Reference 35.

So, in hydrocarbon zones:

$$S_{xo} = \frac{\phi_{EPT}}{\phi_{ND}} \tag{2.36}$$

Figure 2.45 is a quick look at ϕ_{EPT} response compared to FDC, CNL, and induction-log resistivity in gas, oil, fresh water, and saltwater-bearing formations [25]. These responses also indicate moveable oil saturation $(1 - S_{xo})$ and, therefore permeability.

2.3.5 Special Openhole Logs and Services

2.3.5.1 Dipmeter

The dipmeter is a four-armed device with pads that read resistivity of thin zones. These four resistivity curves are analyzed by computer and correlated to determine formation dip and azimuth. The dips are presented on a computer-produced log. In addition to dip, hole deviation, borehole geometry, and fracture identification are also presented.

2.3.5.2 Repeat Formation Tester

The repeat formation tester measures downhole formation pressures. The tool is operated by an electrically driven hydraulic system so that

EPT Quicklook

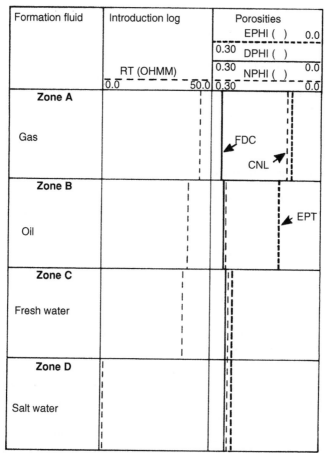

Formation fluid	Introduction log	Porosities

(chart showing Zone A — Gas, Zone B — Oil, Zone C — Fresh water, Zone D — Salt water, with RT (OHMM) 0.0 — 50.0, and porosities EPHI ( ) 0.30 — 0.0, DPHI ( ) 0.30, NPHI ( ) 0.30 — 0.0; curves labeled FDC, CNL, EPT)

FIGURE 2.45 EPT quicklook chart comparing curve response of induction resistivity, neutron porosity, density porosity and EPT porosity in water bearing and hydrocarbon-bearing zones [35].

several zones may be pressure tested on the trip into the hole. Once the drawdown pressure and the pressure buildup have been recorded, they can be processed by a computer at the well-site to provide Horner plots from which permeabilities are calculated. Permeabilities from the drawdown test often vary considerably from measured permeabilities and should be considered an order-of-magnitude estimate. This is usually due to a very shallow depth of investigation associated with drawdown tests [33]. The pressure buildup has a better depth of investigation than the drawdown pressure test. Accuracy depends on what type of pressure-wave propagation model is chosen [33, 35, 36] as well as the compressibility and viscosity of the formation fluids.

2.3.5.3 Sidewall Cores

After drilling, cores from the side of the borehole can be taken by wireline core guns or drills. Guns are less expensive but do not always recover usable cores. Sidewall drilling devices have become quite common in the last few years. Up to 20 cores may be cut and retrieved on one trip into the hole.

Cased Hole Logs   Cased hole logs are run to evaluate reservoir performance, casing/cement jog quality, and to check flow rates from producing intervals. The reader is referred to Bateman's book [38] on cased hole logging which provides a more detailed discussion than is possible in this summary.

Cased hole logs can be broadly divided into two classes:

1. Logs that measure formation parameters through the casing.
2. Logs that measure the parameters within and immediately adjacent to the casing.

These logs are all combined to monitor fluids being produced, monitor reservoir performance, and monitor producing-string deterioration with time. They differ from open-hole logs in that the majority of cased hole logs merely monitor fluid production rather than provide extensive data on formation characteristics.

Following recent developments in cased hole logging technology, the following measurements can be made:

- Formation resistivity through casing: Although the concept of measuring resistivity through casing is not new, but it is only recent break throughs in downhole electronics and electrode design that have made the measurements possible [Schlumberger tool: CHFR].
- Formation porosity through casing: Measurements are based on an electronic neutron source instead of a chemical source, and it uses borehole shielding and focusing to obtain porosity measurements that are affected only minimally by borehole environment, casing standoff, and formation characteristics such as lithology and salinity [Schlumberger tool: CHFP].
- Formation acoustics: This technique measures formation compressional and shear slowness in cased wells [Schlumberger tool: DSI].
- Fluid identification: This tool provides a technique for determining formation pressures in old or new cased wells, and it enables efficient fluid sampling without the inherent risks of conventional or standard sampling techniques [Schlumberger tool: CHDT].

Cased Hole Formation Evaluation   Two tools are currently being used to provide formation evaluation in cased holes:

1. Pulsed neutron logs.
2. Gamma spectroscopy tools (GST) logs.

2.3.5.4 Pulsed Neutron Logs

Pulsed neutron logs are used to monitor changes in fluid content and water saturation with respect to time. Current tools also provide a means of estimating porosity. They are particularly valuable for [38]:

1. Evaluating old wells when old open-hole logs are poor or nonexistent.
2. Monitoring reservoir performance over an extended period of time.
3. Monitoring the progress of secondary and tertiary recovery projects.
4. Formation evaluation through stuck drill pipe (generally a last resort).

2.3.5.5 Theory

A neutron generator that consists of an ion accelerator fires deuterium ions at tritium targets. This produces a burst of 14 keV neutrons which pass through the borehole fluid (must be fresh water), casing, and cement. The burst then forms a cloud of neutrons in the formation which are rapidly reduced to a thermal state by collisions with the atoms in fluids in the rock (made up primarily of hydrogen atoms). Once in a thermal state, they are most liable to be captured by chlorine (or boron). The capture process will produce a gamma ray of capture which is then detected by a scintillometer in the tool. The time it takes for the neutron cloud to die during the capture process is a function of the chlorine concentration in the formation fluid. This is then related to water saturation. Rapid disappearance of the thermal neutron cloud indicates high water saturation. Slower disappearance of the cloud indicates low water saturation (i.e., high hydrocarbons saturations). The rate of cloud decay is exponential and can be expressed by:

$$N = N_o e^{(-t/\tau)} \qquad (2.37)$$

where N = number of gamma rays observed at time t
N_o = number of gamma rays observed at $t = 0$
t = elapsed time (microseconds)
τ = time constant of the decay process, microseconds

Of most interest is τ since it is strictly a function of the decay rate of the neutron cloud (or rather the slope of the exponential function). From τ the capture cross section, Σ, can be calculated:

$$\Sigma = 4,550/\tau \qquad (2.38)$$

The tools that are available to measure τ include:

1. TDT-K (with 3 moveable gates or detectors)
2. DNLL (dual neutron lifetime log) (which uses 2 gates),
3. TDT M (with 16 fixed gates), and
4. TMD (thermal multigate decay) (which uses 6 gates).

In general, these tools all perform the same function: they measure the decay rate of the neutron cloud in the formation. This is accomplished by using a series of windows to measure near and far-spacing counting rates, as well as background gamma ray rates. The first gates are not triggered until all neutrons in the cloud in the formation are thermalized. At this point neutron capture has started. By using certain gating times and gate combinations, the slope of the straight portion of the decay curve is measured and related to Σ. In addition, the ratio between the short-space-detector and long-space-detector counting rates is also calculated and is related to porosity. (It is similar in principle to the CNL porosity device used in openhole logging).

2.3.5.6 Log Presentations

Figure 2.46 is an example of a DNLL log presentation. Most other TDT logs are presented in a similar way, except that the number of curves varies from company to company. The log shown in this figure consists of 5 curves:

1. Gamma ray curve (Track 1).
2. Gate 1 counting rate (CPM).
3. Gate 2 counting rate (CPM).
4. Ratio curve ($\approx$ CNI. ratio).
5. Sigma curve (Σ).

In addition, the pips located on the left side of track 2 are the corrected casing collar locations.

Gate 1 and Gate 2 show the raw data from the detectors, the ratio curve shows relative hydrogen concentration ($\approx$ water-filled porosity), and the σ curve shows the capture cross-section. Some logs also show a τ curve, but it is normally omitted [38].

2.3.5.7 Interpretation

Interpretation of pulsed neutron logs is very straightforward. It relies on knowledge of three parameters (four in hydrocarbon-bearing zones):

1. Σ_{log} (capture units).
2. Σ_{matrix}
3. Σ_{water}
4. $\Sigma_{hydrocarbon}$

According to Schlumberger [41] the log response may be described as:

$$\Sigma_{log} = \Sigma_{ma}(1-\phi) + \Sigma_w \phi S_w + \Sigma_{hy}\phi(1-S_w) \qquad (2.39)$$

Solving for S_w:

$$S_w = \frac{(\Sigma_{log} - \Sigma_{ma}) - \phi(\Sigma_{hy} - \Sigma_{ma})}{\phi(\Sigma_w - \Sigma_{hy})} \qquad (2.40)$$

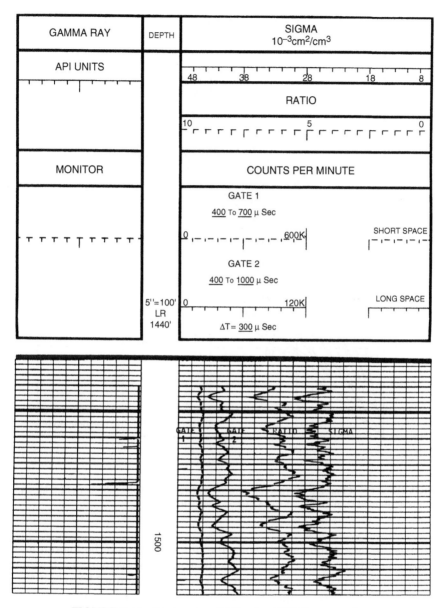

FIGURE 2.46 Dual neutron lifetime log (DNLL) presentation.

Porosity (ϕ) can be found either from an openhole porosity log or by combining the ratio curve and $\Sigma_{\log}$. Figures 2.47 to 2.50 are charts to find porosity, apparent water salinity, and Σ_w, using this combination. By selecting the chart for the appropriate borehole and casing diameter, these values

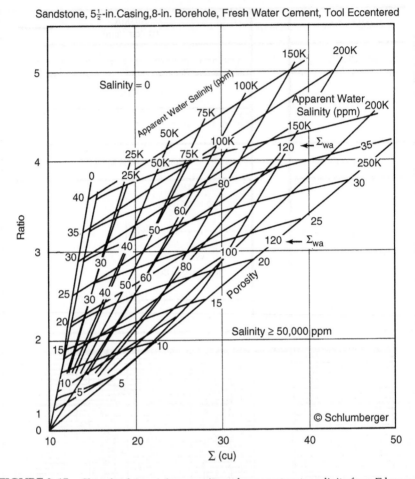

FIGURE 2.47 Chart for determining porosity and apparent water salinity from Σ log and ratio curves in 51/2-in. casing and an 8-in. borehole [20].

are easy to determine. Simply enter the proper axes with the log-derived values, and find porosity and Σ_w at the intersection of the two lines. If water salinity and formation temperature are known, use Figure 2.50 to find Σ_w.

Estimating Σ_{hy} is another matter. You must first know if the hydrocarbons are oil, methane, or heavier hydrocarbon gases (i.e., propane, butane, pentane). For oil, solution gas-oil ratio and oil gravity (° API) are needed. If the gas is methane, reservoir pressure and temperature are required. For gases other than methane, the specific gravity of the gas can be converted to equivalent methane using Figure 2.51, and then Figure 2.52 can be entered. Once all of the parameters have been found, Equation 2.40 is used to find S_w.

Sandstone, 7-in. Casing, 10-in. Borehole, Fresh Water Cement, Tool Eccentered

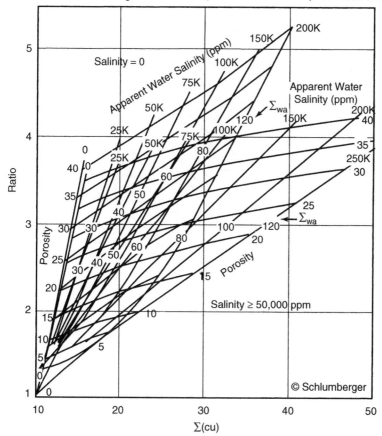

Example: Ratio = 3.1
Σ_{LOG} = 20 cu
Borehole fluid salinity = 80,000 ppm
$5\frac{1}{2}$-in. casing cemented in $8\frac{3}{4}$-in. borehole

Thus, from Chart Tcor-3
ϕ = 30%
Apparent water salinity = 50,000 ppm
Σ_{wa} = 40 cu

If this were a clean formation and connate water salinity was known to be 150,000 ppm, then

$$S_w = \frac{50,000}{150,000} = 33\%$$

FIGURE 2.48 Chart for determining porosity and apparent water salinity from Σ log and ratio curves in 7-in. casing and a 9-in. borehole [20].

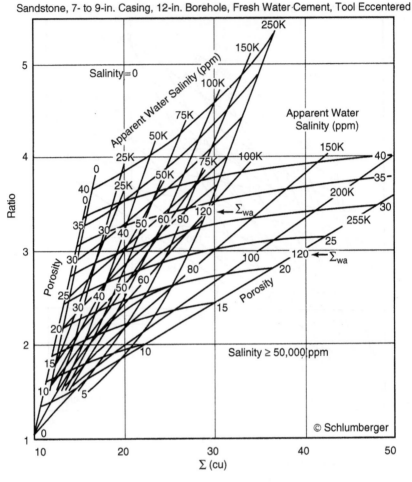

Sandstone, 7- to 9-in. Casing, 12-in. Borehole, Fresh Water Cement, Tool Eccentered

FIGURE 2.49   Chart for determining porosity and apparent water salinity from $\sum$ log and ratio curves in 7- to 9-in. casing and a 12-in. borehole [20].

2.3.5.8 *Applications*

Pulsed neutron logs are most useful for monitoring changes in water saturation over time while a reservoir is produced. Initially, these logs are run prior to perforating a zone. Subsequent logs are run every few months (or years) depending on production rate and the amount of control desired. Water saturation is calculated for each run using Equation 2.40 and subtracted from saturations determined from earlier runs. These values (ΔS_w) show the change in the position of the water table (hydrocarbon-water contact) versus time.

Another application is monitoring residual oil saturation in waterflood projects. The procedure outlined by Bateman [38] involves first running a

Example: A reservoir section at 90°C temperature and 25 MPa pressure contains water of 175,000 ppm (NaCl) salinity: 30°API oil with a GOR of 2000 cu ft/bbl, and methane gas.

Thus, $\Sigma_w = 87$ cu ($\tau = 52$ µs)

$\Sigma_0 = 19$ cu ($\tau = 240$ µs)

$\Sigma_g = 6.9$ cu ($\tau = 660$ µs)

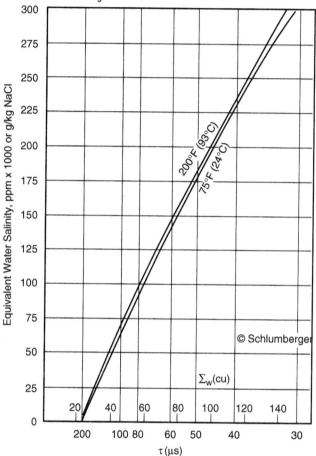

FIGURE 2.50 Chart for finding Σ_w from equivalent water salinity and formation temperature [20].

base log (prior to injection). Next, salt water is injected and another log is run. Then fresh water is injected and another log run. If Σ_{brine} and Σ_{fresh} are known, Bateman [38] suggests using:

$$S_o = 1 - \frac{\Sigma_{log(brine)} - \Sigma_{log(fresh)}}{\phi(\Sigma_{brine} - \Sigma_{fresh})} \qquad (2.41)$$

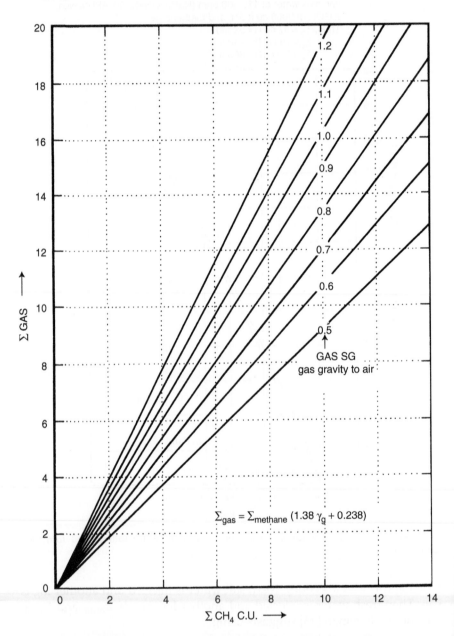

FIGURE 2.51   Chart for converting $\sum$gas to $\sum$methane [38].

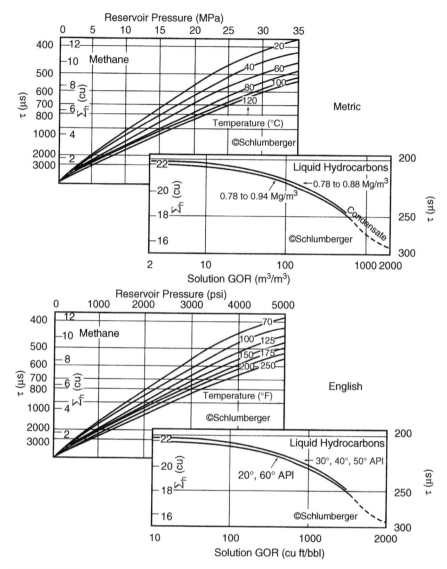

FIGURE 2.52 Charts for finding Σh (for methane and liquid hydrocarbons) from reservoir pressure and formation temperature (gas) or solution gas-oil ratio (GOR) and API gravity (liquid 7 hydrocarbons) [20].

to find residual oil saturation (S_{or}) Additional details of estimating S_{or} are given later.

The main problem with using these logs is the presence of shale. Shale normally appears wet, and shale will make a reservoir look like it has higher S_w. Openhole logs and an NGS log are needed to confirm this interpretation although shaly sand corrections can easily be made [38].

2.3.5.9 Gamma Spectroscopy Tools (GST)

Also known as the carbon-oxygen log, this device has recently been incorporated into pulsed neutron tools to aid in differentiating oil and gas from water. GST tools operate with the same neutron generator as the pulsed neutron devices, but gamma rays returning from the formation are measured.

Two types of gamma rays are produced when neutrons are fired into a formation:

1. Those that result from neutron capture by chlorine and boron.
2. Those that result from inelastic collisions with atoms.

The detector on this tool has energy windows set to receive certain returning gamma rays [39]. The detectors are protected from the fast neutron source by an iron shield, and from returning thermal neutrons by a boron shield.

The energy of the returning gamma rays depends on the atom involved in the collisions. The atoms of interest include carbon, oxygen, silicon, and calcium. Carbon-oxygen ratio is a carbon indicator and when combined with porosity, gives an estimate of water saturation if matrix lithology is known. Figure 2.53 is used for this determination. Silicon-calcium ratio is an indicator of matrix and is used to distinguish oil-bearing rock from calcareous rocks (such as limy sands and limestones) [39, 40]. Figure 2.54 is an example of a carbon-oxygen log.

If capture gamma rays are also detected with separate energy windows, chlorine and hydrogen content can be determined and related to formation water salinity. Figures 2.55 and 2.56 are used for this purpose. All that is required to estimate salinity of formation waters is knowledge of borehole fluid salinity, Cl/H ratio, and response mode of the tool. These devices should *not* be confused with the natural gamma spectroscopy log which only measures naturally-occurring gamma rays.

The readings on the GST log are not affected by shale although carbonaceous shales can cause trouble because of the sensitivity to carbon. Usually, however, these effects can be calibrated for or taken into account when this log is interpreted. Much of the interpretation of this type of log is based in regional experience; the analyst should have a good idea of the types of rocks present before trying to make an interpretation. No lithology cross-plot charts are presently available to estimate lithology with these logs.

2.3.5.10 Natural Gamma Spectroscopy

This log operates in the same manner as its openhole counterpart. The main difference is that the log should be calibrated prior to being run in

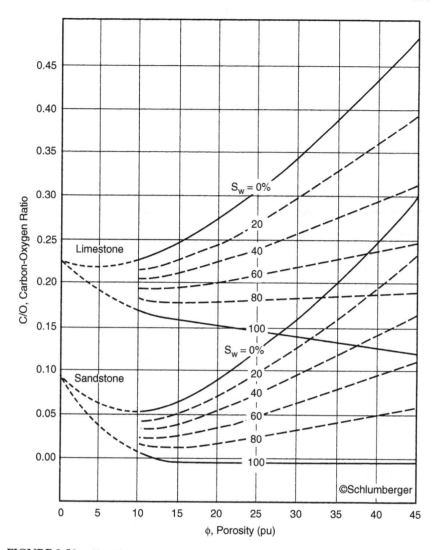

FIGURE 2.53 Chart for finding water saturation using the carbon-oxygen ratio curve and porosity if matrix lithology is known [20].

cased holes. No correction charts are currently available for cased hole applications with this device. Curve presentations are the same as for the open-hole version. Refer to the open-hole section for a discussion of this log.

Cased-Hole Completion Tools These tools examine cement bond and casing quality. They assure that no leakage or intercommunication will

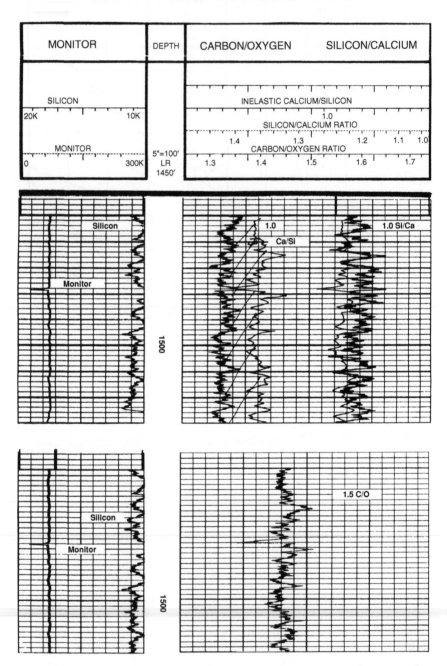

FIGURE 2.54 Carbon-oxygen and silicon-calcium ratio curves on a carbon-oxygen log.

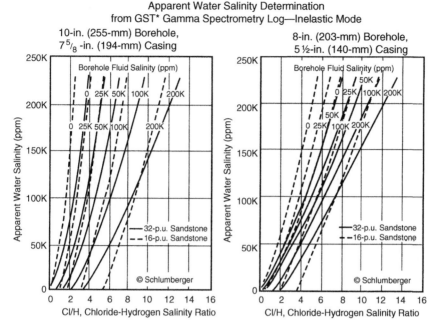

FIGURE 2.55 Chart for finding apparent water salinity from chlorine-hydrogen ratio and borehole fluid salinity from a GST log (in elastic mode) [20].

occur between producing horizons, or between water-bearing horizons and producing horizons. The most common completion tools include:

1. Cement bond logs (CBL).
2. Multifingered caliper logs.
3. Electromagnetic inspection logs.
4. Electrical potential logs.
5. Borehole televiewers.

2.3.5.11 Cement Bond Logs (CBL-VDL)

Cement bond logs are used to check cement bond quality behind the casing and to estimate compressive strength of the cement. It can also be used to locate channeling in the cement or eccentered pipe and to check for microannulus.

2.3.5.12 Theory

The cement bond tool is the same as a conventional sonic tool except that the receiver spacings are much larger. It consists of a transmitter and two receivers. The near receiver is 3 ft below the transmitter and is used to find Δt for the casing. The far receiver is 5 ft below the transmitter and is used for the variable density log(VDL) sonic-wave-form output.

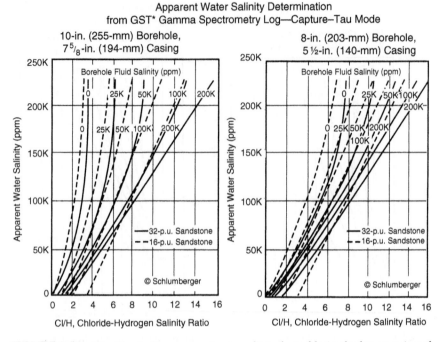

FIGURE 2.56   Chart for finding apparent water salinity from chlorine-hydrogen ratio and borehole fluid salinity derived from a GST log (Tau-capture mode) [20].

The operation is the same as a conventional sonic except that the transmit time (one way) is measured. The transmitter is fired and a timer is triggered in both receivers. The wave passes through the fluid in the casing, the casing, and the cement, and into the formation. The near receiver measures the first arrival of the compressional wave and the timer is shut off. This Δt is a function of whether the casing has cement behind it or not.

The sound wave is then picked up by the lower receiver which recognizes refracted compressional wave arrivals from the casing, cement, and formation, as well as Rayleigh, Stonely, and mud-wave arrivals. Figure 2.57 shows the basic tool configurations.

The most important parameter measured by this tool is compressive-wave attenuation-rate. This parameter is a function of the amount of cement present between the pipe and formation. Typically, cement must be at least $\frac{3}{4}$ in. thick on the casing in order for attenuation to be constant [38]. Each part of the log reads different attenuations. The CBL registers attenuation of the compressional wave in the cement and casing which gives an indication of the cement-casing bond-quality. The VDL registers the attenuation of the compressional wave through casing, cement, and formation which gives an indication of acoustic coupling between casing, cement, and surrounding

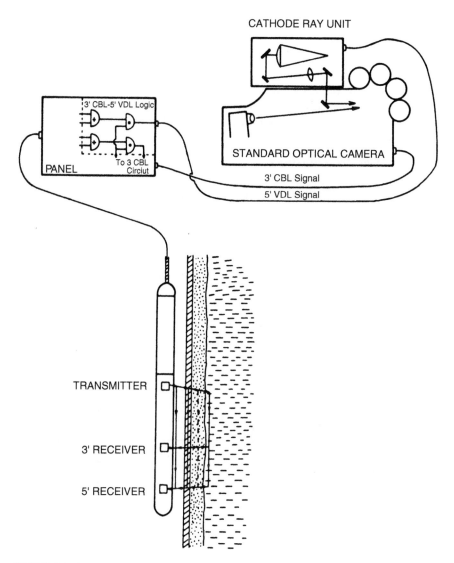

FIGURE 2.57 Transmitter-receiver arrangement and surface equipment for cement bond log-variable density log combination [41].

rock. This indicates not only the casing bond quality but also the cement-formation-bond quality.

The basic parameter used to evaluate cement bonds is called the *bond index* and can be calculated by

$$\text{bond index} = \frac{\text{attention in zone of interest (db/ft)}}{\text{attenuation in a well cemented section (db/ft)}} \qquad (2.42)$$

Bond index gives a relative way to determine bond quality through any given section of pipe. The minimum value of bond index necessary for a good hydraulic seal varies from region to region, and depends on hole conditions and type of cement used. Ideally, an index of 1.0 indicates excellent pipe-cement bonding; decreasing values show deteriorating conditions which may require squeezing to bring bonding up to acceptable standards. A bond index curve may be presented in track 2.

2.3.5.13 Log Presentations

Figure 2.58 shows a CBL-VDL log. Typically, three curves (and sometimes more, depending on the service company) are Presented on the log. Track I contains total travel time. This is total one-way travel time and is a function of the casing size and tool centering. Other curves may be presented in this track, including gamma ray, neutron, and casing collar locator logs. Track 2 contains the cement bond logs amplitude curve. The log is scaled in millivolts and is proportional to the attenuation

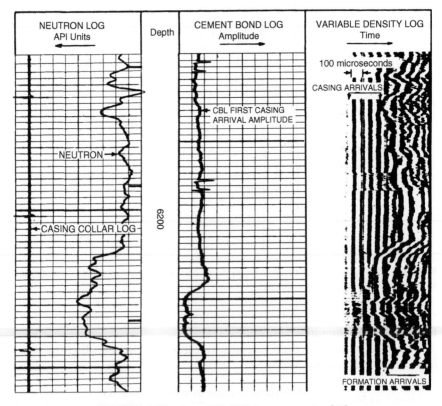

FIGURE 2.58   Basic CBL-VDL log presentation [41].

of the compressional-wave first-arrivals. High attenuation produces low-amplitude values; low attenuation produces high-amplitude values. The higher the amplitude, the poorer is the casing-to-cement bond. Direction of increasing amplitude is normally indicated by an arrow. Some presentations also include a bond index curve in track 2. Track 3 contains the variable density log (VDL) display. The most common presentation is dark- and light-colored bands that represent the peaks and valleys of the wave train. Figure 2.58 shows two possible types of arrivals:

1. Those from the casing which appear as straight, parallel light and dark bonds intermittently broken by small V-shaped spikes which indicate the position of casing collars.
2. Those from the formation which appear as wavy, irregular, and inter-mittent light and dark bands which represent curve attenuation in the rock surrounding the borehole.

2.3.5.14 Interpretation

Interpretation of CBL-VDL logs involves recognition of basic curve pat-tern for determining whether casing is properly bonded or not. These curve patterns are presented in Figures 2.59 to 2.62. Four basic types of patterns are apparent:

1. Those that show strong casing arrivals only.
2. Those that show strong casing and formation arrivals.
3. Those that show weak casing arrivals and strong formation arrivals.
4. Those that show both weak casing and weak formation arrivals.

Strong casing arrivals are shown in Figure 2.59 and are characterized by the pronounced casing arrival pattern (straight, alternating light and dark bands). No formation arrivals are present and cement-bond log-amplitude is moderate to high. These indicate free pipe with no cement or cement-casing-formation coupling. A high amplitude curve reading indicates low attenuation, hence no cement in the annulus.

Strong casing and formation arrivals are shown in Figure 2.60. This pat-tern has both the clean, pronounced casing signature as well as a strong, wavy-formation signature. The lack of cement is indicated by the high cement bond log-amplitude (i.e., no cement attenuation). The combination of these signals is interpreted as eccentered casing in contact with the wall of the well-bore. In this situation, proper cementation may be impossible.

Weak-casing and strong-formation arrivals are shown in Figure 2.61. This pattern shows no apparent casing or very weak casing patterns and very strong formation-patterns nearly filling the VDL in track 3. This indi-cates good casing-cement formation bonding, confirmed by the low cement bond log amplitude (high attenuation). Rayleigh and mud wave arrivals are also apparent along the right side of track 3.

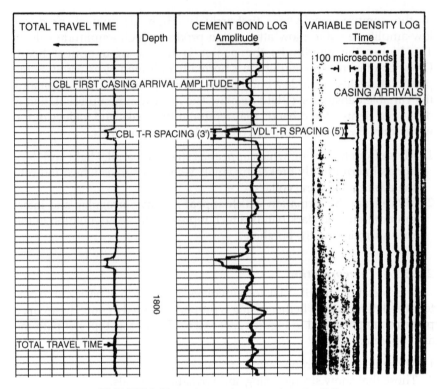

FIGURE 2.59   CBL-VDL log run in free pipe [41].

Weak-casing and weak-formation arrivals are shown in Figure 2.62. This pattern shows what appears to be a slightly attenuated casing pattern but the cement bond log amplitude suggests otherwise. The curve indicates strong to very strong attenuation due to cement. Strongest attenuation occurs at "A" with a very weak formation pattern on the VDL. Comparison with the open-hole VDL (immediately to the right) shows no unusual attenuations of the formation signal. This also confirms poor acoustic coupling between cement and formation with good coupling between casing and cement.

Other possible interpretations for this type of pattern are possible.

1. Gas in the mud can be ruled out by examining long intervals of the log. Generally, this effect will occur over long rather than short sections in the well.
2. Eccentered tool in the casing, which causes destructive interference of compressive-wave first-arrivals, can be confirmed by checking for wiggly casing arrivals or a slight decrease in the casing-arrival time shown on the total-transit-time curve or VDL log [41].

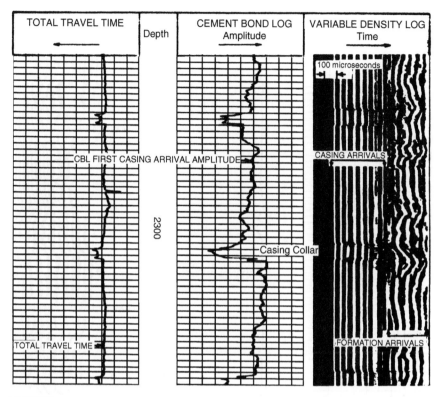

TOTAL TRAVEL TIME	Depth	CEMENT BOND LOG Amplitude	VARIABLE DENSITY LOG Time

FIGURE 2.60 CBL-VDL log run in casing eccentered in borehole making contact with the formation [41].

3. Thin cement sheaths, caused by excessive mud cake thickness along a permeable formation, are a problem when cement sheaths are less than $\frac{3}{4}$ in. thick (which allows stronger casing arrivals). At times, the cement and formation have a slight acoustic coupling which gives the VDL a faint or weak formation signal.

2.3.5.15 Microannulus or Channeling in Cement

Microannulus occurs when the cement is emplaced and the casing is pressurized. When pressure is released after the cement has set, the casing "pops" away from the cement sheath. This generates a gap or microannulus between the casing and the cement. This can also occur if excessive pipe dope or varnish is present on the pipe. Microannulus due to pressurization primarily occurs opposite washed out portions of the borehole.

Channeling occurs when cement is in the annulus but does not completely surround, or is not bonded to, the pipe. This condition will not have proper fluid seal which allows oil, gas or water to pass up the hole

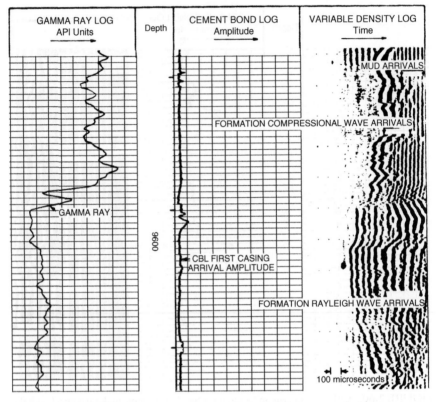

FIGURE 2.61 CBL-VDL log run in well-bonded casing [41].

outside the casing. Microannulus, on the other hand, may have a proper seal even though a small gap exists. It is very important to be able to distinguish between microannulus and channeling; squeezing may eliminate the channel altogether.

Figure 2.63 shows a case of microannulus. Figure 2.63a shows strong to weak casing-bond on the CBL amplitude and weak formation-arrivals on the VDL. This indicates poor acoustic coupling between casing, cement, and formation. The weaker the formation signal, the more pronounced the microannulus.

Microannulus can be easily differentiated from channeling by:

1. Pressurizing the pipe and rerunning the CBI-VDL. Microannulus conditions are confirmed by strengthened formation signals on the VDL and decreased CBL amplitude indicating better casing-cement acoustic coupling. Typically, channeling will produce little or no improvement in the signal when the casing is pressurized. Figure 2.63b shows a case of stronger formation arrivals indicating the presence of microannulus rather than channeling.

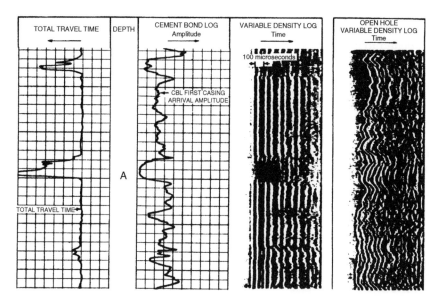

FIGURE 2.62 CBL-VDL log run in casing with a good cement-casing bond and a poor cement-formation bond [41].

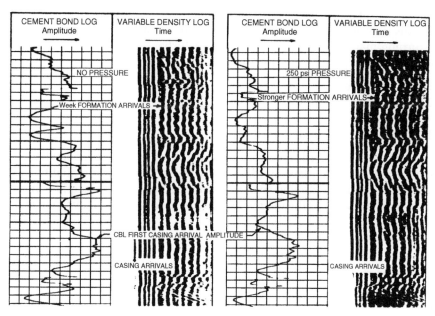

FIGURE 2.63 CBL-VDL log showing effects of microannulus and the change in signal strength on VDL in pressured and nonpressured pipe [41].

2. Microannulus tends to occur over long intervals of the log; channeling is a localized phenomena. This is a result of microannulus being directly related to pipe expansion during cementing operations.

Once the log has been interpreted, remedial measures can be applied as necessary.

2.3.5.16 Multifingered Caliper Logs

These logs incorporate up to 64 feelers or scratchers to examine pipe conditions inside the casing. Specifically, they can be combined with other logs to check:

1. Casing collar locations.
2. Corroded sections of pipe.
3. Casing wear.
4. Casing cracks or burstings.
5. Collapsed or crushed casing.
6. Perforations.
7. Miscellaneous breaks.

The number of feelers is a function of pipe diameter; smaller diameter pipe requires fewer feelers on the tool.

2.3.5.17 Electromagnetic Inspection Logs

This device induces a magnetic field into the casing and measures the returning magnetic flux. In general any disturbance in the flux from readings in normal pipe can be used to find:

1. Casing collars.
2. Areas of corroded pipe.
3. Perforations.
4. Breaks or cracks in the pipe.

This tool only records if corrosion has occurred on the pipe, not whether it is currently taking place. It does give an indication of casing quality and integrity without removing the pipe from the hole. The principle behind this tool is the same as the magna flux device used to detect flaws in metals in a machine shop.

2.3.5.18 Electrical Potential Logs

Similar in some respects to an SP log, this tool measures the potential gradient of a DC current circulating through a string of casing. This current is applied to provide the casing with cathodic protection thereby preventing casing corrosion; any deviation from a negative field suggests that the pipe is not receiving proper protection and is probably being corroded.

Combined with an electromagnetic inspection log, areas currently undergoing corrosion as well as having a relative amount of damage can be determined with ease.

2.3.5.19 Borehole Televiewers

This tool incorporates an array of transmitters and receivers to scan the inside of the casing. The signals are sent to the surface where they are analyzed and recorded in a format that gives a picture of the inside of the casing. Any irregularities or cracks in the pipe are clearly visible on the log presentation. This allows engineers to fully scan older pipe and get an idea of the kind and extent of damage that might not otherwise be readable from multifinger caliper, electromagnetic inspection, or electrical potential logs. The main drawback to this device is that it must be run in a liquid-filled hole to be effective.

Production Logs   Production logs are those devices used to measure the nature and behaviour of fluids in a well during production or injection. A Schlumberger manual [42] summarizes the potential benefits of this information:

1. Early evaluation of completion efficiency.
2. Early detection of disturbances which are not revealed by surface measurements (thief zones, channeled cement, plugged perforations, etc.).
3. Detailed information on which zones are producing or accepting fluid.
4. More positive monitoring of reservoir production.
5. Positive identification of encroachment, breakthrough, coring, and mechanical leaks.
6. Positive evaluation of injection efficiency.
7. Essential guidance for remedial workover and secondary or tertiary recovery projects.

The reader is referred to the Schlumberger volume [42] on production log interpretation for examples of various cased-hole-log situations. It is still free upon request.

The types of logs run include

1. Temperature
2. Manometer and gradiomanometer
3. Flow meters
4. Radioactive tracers

Devices that measure water-holdup are also available. These logs can be run singly or in combination on a production combination tool so that a number of parameters may be recorded on the same log sheet

2.3.5.20 Temperature Log

A thermometer is used to log temperature anomalies produced by the flow or fluid inside the casing or in the casing annulus. It is used to help determine flow-rates and points of fluid entry or exit, and is, perhaps, most useful for finding fluid movement behind the casing.

2.3.5.21 Injection Wells

Figure 2.64 is the response of the temperature log when fluid is being injected into a reservoir. The sloping portion defines the geothermal gradient the vertical portion defines the zone tasking the water and is a function of the geothermal gradient as well as the injection fluid temperature. Below the sloping position, the temperature/curve rapidly returns to normal formation temperature and the geothermal gradient. The vertical portion of the log clearly indicates where the fluid is leaving the casing.

2.3.5.22 Production Wells

Figure 2.65 is the response of the temperature log when fluid is flowing into a well from perforations in the casing. Three curves are presented. This figure shows that curve response depends on whether the fluid produced is hotter, the same as, or cooler than the geothermal gradient. If the fluid is hotter or cooler, then the entry point is obvious. If the fluid temperature is the same as the geothermal gradient, the change is so subtle that recognition of the entry point may be very difficult. In this case, a high resolution thermometer may be necessary to pinpoint the fluid entry location.

2.3.5.23 Flow Behind Casing (Annular Flow)

Figure 2.66 is a typical response to annular flow down the outside of the casing in a shut-in well. The figure shows water entering the annulus at about 6,500 ft. Perforations are at ∼ 8,500 ft.

In a producing well, the shape of the curve defines the top of the annular space and its relationship to the perforations.

2.3.5.24 Manometers and Gradiomanometers

Manometers are pressure-sensitive devices used to measure changes in pressure that result from:

1. Leaks in tubing or casing.
2. Fluid inflow through perforations.
3. Gradient measurements in a static mud column.

They are particularly useful for determining pressure opposite a gas-bearing horizon. This value is vital for calculating open-flow potentials in gas wells.

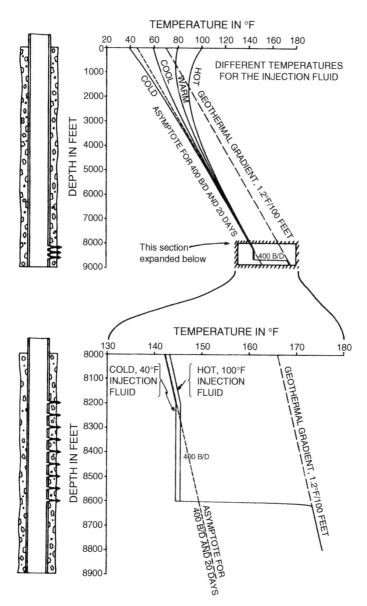

FIGURE 2.64   The effect of water injection on a temperature log for several injection water temperatures [42].

Gradiomanometers are used to check the difference in pressure over a 2-in. interval in a producing well. This is then related to water-holdup in polyphase fluid flow within the casing.

The pressure difference is converted to density and is used to interpret two-phase flow (usually consisting of water as the heavy component and

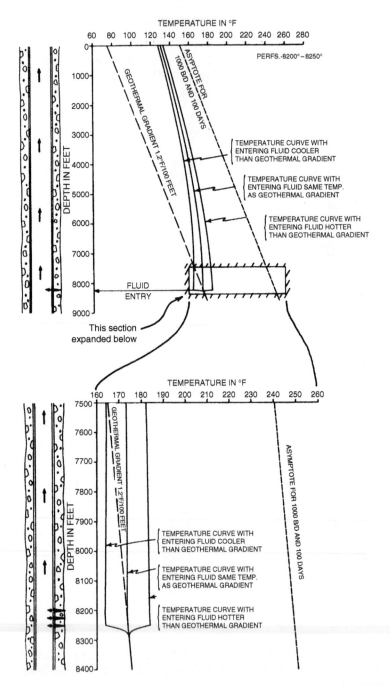

FIGURE 2.65   Temperature logs in a producing interval for formation fluids with different temperatures [42].

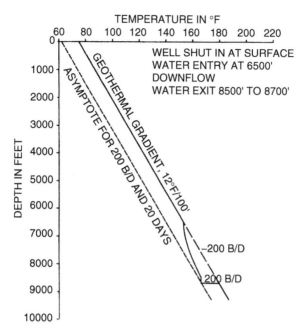

FIGURE 2.66 Temperature log showing water flow behind the casing. Water is flowing down to the producing interval [42].

oil as the lighter component). At any given level, the gradiomanometer measures the specific gravity (density) of any fluids entering the borehole. The log reading is related to water holdup and specific gravity by:

$$\rho_{\text{gradiomanometer}} = Y_w \rho_w + (1 - Y_w)\rho_o \tag{2.43}$$

where $\rho_{\text{gradiomanometer}} =$ specific gravity reading of the gradiomanometer, g/cc
$\rho_w =$ specific gravity of the formation water, g/cc

$\rho_o =$ specific gravity of the oil being produced with the water, g/cc
$Y_w =$ water holdup (or holdup of the heavy phase)

The specific gravity reading is not exclusively dependent on fluid density. Since the fluids are also flowing while measurements are being made, other terms must be added:

$$\rho_{\text{gradiomanometer}} = \rho_f(1.0 + K + F) \tag{2.44}$$

where $\rho_{gradiomanometer}$ = specific gravity reading of the
                 instrument, g/cc
                 ρ_f = specific gravity of the fluid in the
                 casing (oil + water + gas), g/cc
                 K = a kinetic term
                 F = a friction term from fluid flowing
                 around the tool

At flow rates less than 2,000 bopd, F is negligible and $\rho_{grad} \approx \rho_f$ [42]. The kinetic term is important when logging from tubing into the casing. Fluid velocity changes become significant at the change in hole diameter. The change causes a sharp pressure increase on the log. The friction term is important at high flow rates in casing and when logging in small-diameter tubing or casing [42].

Another log similar to the gradiomanometer is the water-holdup meter. The main limitation is that it only reads water holdup and cannot be used if water is not present. It has the advantage over the gradiomanometer where sensitivity is required; small differences in density may not be seen by the gradiomanometer.

2.3.5.25 Flowmeters

Flowmeters are designed to measure fluid flow in the casing. This measurement is then related to volume of fluid being produced. Three types of flowmeters are available:

1. Fullbore-spinner flowmeter.
2. Continuous flowmeter.
3. Packer flowmeter.

Each is used in certain circumstances and is combinable with other devices so that improved flow rates can be obtained.

2.3.5.26 Fullbore-Spinner Flowmeter

This device measures velocities of fluid moving up the casing. These velocities are then related to volume of fluid moved with charts available from the various service companies. In general, this tool can be used at flow rates as low as 20 barrels per day in monophase flow situations (usually water). Polyphase flow raises this minimum to 300 barrels per day if gas is present (i.e., oil and water) in $5\frac{1}{2}$ in. casing. This tool is used in wells with hole diameters ranging from $5\frac{1}{2}$ to $9\frac{5}{8}$ in.

2.3.5.27 Continuous Flowmeter

This tool is similar to the fullbore-spinner flowmeter except that it can be applied to hole diameters between $3\frac{1}{2}$ in. and $6\frac{5}{8}$ in. It has a higher flow threshold (in barrels per day) and should be restricted to use in monophase flow situations (i.e., waterfloods, high-flow-rate gas wells, and high-flow-rate oil wells) [42]. It can be combined with a spinner flowmeter for better flow measurements.

2.3.5.28 Packer Flowmeter

This is a small spinner-flowmeter with an inflatable packer that can be used in small-diameter tubing ($1^{11}/_{16}$ to 2^{1}_{8} in.) It has an operable flow range from 10 to 1,900 barrels per day an can be applied in low-flow wells as long as measurements are made in the tubing at a sufficient distance above the perforations. Flow measurements are related to volume of fluid flowing the same way found with the other spinner flowmeters.

2.3.5.29 Radioactive Tracers

Radioactive tracers are combined with cased hole gammaray logs to monitor:

1. Fluid velocities in monophase fluid flow situations where flow velocity is at or near the threshold for spinner flowmeters.
2. Fluid movement behind the casing or to locate channeling in the cement.

Fluid velocity is measured by velocity-shot analysis. A shot of radioactive fluid is injected into the flow stream above two detectors located on a stationary mammary tool. As the radioactive pulse moves down the hole, the amount of time required to move past the two detectors is measured. This travel time is then related to flow rate in the casing by:

$$q(B/D) = \frac{\text{spacing(in.)} \times \frac{1(ft)}{12(in.)} \times \frac{\pi}{4}(d_h - d_{tool})(in.^2) \times \frac{1(ft.^2)}{144(in.^2)} \times 256.5 \frac{B/D}{ft^3/min}}{\text{time(sec)} \times \frac{1(min)}{60(sec)}} \qquad (2.45)$$

where q is flow rate in barrels per day, the spacing between detectors is in inches, the time between detector responses is in seconds, d_h is the hole diameter in in., and d_{tool} is the tool diameter in in.

The main limitation is that slippage and water-holdup factors seriously affect the time reading so this technique cannot be applied in production wells. Moreover, the production of radioactive material is not desirable; therefore, use is mainly restricted to water- or gas-injection wells [42].

Fluid movement behind the casing can be measured with a timed-run radio-active survey. A slug of radioactive fluid is introduced at the bottom

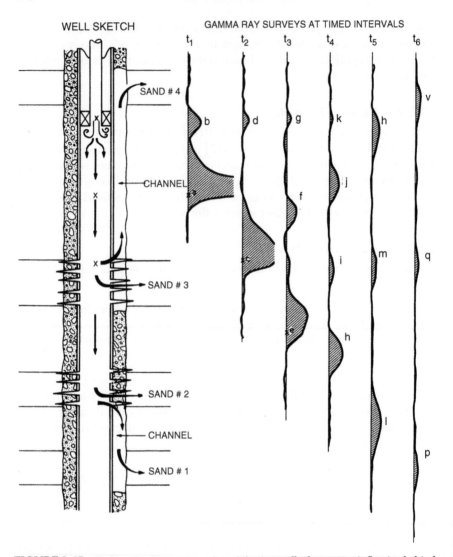

FIGURE 2.67 Radioactive tracer survey in an injection well where water is flowing behind the casing into another zone [42].

of the tubing, and movement is then monitored by successive gamma-ray log runs. Unwanted flow up any channels in the cement can be easily determined and remedial action taken. Again, this technique is mainly applied to water injection wells to monitor flood operations and injection-fluid losses. Figure 2.67 is an example of this type of application.

2.4 DETERMINATION OF INITIAL OIL & GAS IN PLACE

2.4.1 Initial Oil in Place

For undersaturated crude, the reservoir contains only connate water and oil with their respective solution gas contents. The initial or original oil in place can be estimated from the volumetric equation:

$$N = \frac{7{,}758\,V_b\phi(S_{oi})}{B_{oi}} = \frac{7{,}758\,Ah\phi(1 - S_{wi})}{B_{oi}} \tag{2.46}$$

The constant 7,758 is the number of barrels in each acre-ft, V_b is bulk volume in acre-ft, ϕ is the porosity (ϕV_b is pore volume), S_{oi} is the initial oil saturation, B_{oi} is the initial oil formation volume factor in reservoir barrels per stock tank barrel, A is area in ft^2, h is reservoir thickness in ft, and S_{wi} is the initial water saturation.

In addition to the uncertainty in determining the initial water saturation, the primary difficulty encountered in using the volumetric equation is assigning the appropriate porosity-feet, particularly in thick reservoirs with numerous non-productive intervals. One method is to prepare contour maps of porosity-feet that are then used to obtain areal extent. Another method is to prepare isopach maps of thickness and porosity from which average values of each can be obtained. Since recovery of the initial oil can only occur from permeable zones, a permeability cutoff is used to obtain the net reservoir thickness. Intervals with permeabilities lower than the cutoff value are assumed to be nonproductive. The absolute value of the cutoff will depend on the average or maximum permeability, and can depend on the relationship between permeability and water saturation. A correlation between porosity and permeability is often used to determine a porosity cutoff. In cases in which reservoir cores have been analyzed, the net pay can be obtained directly from the permeability data. When only logs are available, permeability will not be known; therefore a porosity cutoff is used to select net pay. These procedures can be acceptable when a definite relationship exists between porosity and permeability. However, in very heterogeneous reservoirs (such as some carbonates), estimates of initial oil in place can be in error. A technique [43] has been proposed in which actual pay was defined using all core samples above a specific permeability cutoff and apparent pay was defined using all core samples above a specific porosity cutoff; the relationship between these values was used to find a porosity cutoff.

2.4.2 Initial Gas in Place

For the foregoing case of an undersaturated oil (at the bubble point with no free gas), the gas in solution with the oil is:

$$G = \frac{7{,}758 \, Ah\phi(1-S_{wi})R_s}{B_{oi}} \qquad (2.47)$$

where G is the initial gas in solution in standard cubic feet (scf), R_s is gas solubility in the oil or solution gas-oil ratio (dimensionless), and the other terms are as defined in Equation 2.46.

2.4.3 Free Gas in Place

Free gas within a reservoir or a gas cap when no residual oil is present can be estimated:

$$G = \frac{7{,}758 \, V_g\phi(1-S_{wi})}{B_{gi}} \qquad (2.48)$$

where 7,758 is the number of barrels per acre-ft, V_g is the pore volume assigned to the gas-saturated portion of the reservoir in acre-ft, B_{gi} is the initial gas formation volume factor in RB/scf, and the other terms are as already defined. (Note: If the formation volume factor is expressed in ft^2/scf, 7,758 should be replaced with $43{,}560 \, ft^3/acre\text{-}ft$).

2.5 PRODUCTIVITY INDEX

The productivity index, J, is a measure of the ability of a well to produce hydrocarbon liquids:

$$J_o = \frac{q_o}{p_e - p_{wf}} \qquad (2.49)$$

where q_o is the flow rate of oil in stock-tank barrels of oil per day, p_e is the external pressure in psi, P_{wf} is the flowing bottomhole pressure in psi, and the quantity $(p_e - p_{wf})$ is referred to as the pressure drawdown. Because the flow rate in this case is in STB/D, the oil productivity index (J_o) has units of STB/D/psi. Since only q and p_{wf} can be measured directly, p_e is commonly replaced with $\bar{p}$ which can be determined from pressure transient testing.

After the well has been shut in for a period of time (usually at least 24 to 72 hours or longer depending on reservoir characteristics), the well is put on production at a low rate with a small choke. The rate of production is recorded as a function of flow time. When the production rate has stabilized, the flow rate is increased by increasing the choke, and flow rate is monitored with time. This process is repeated until a series of measurements has been recorded [19].

In order to attain a stabilized productivity index, a minimum time is required after each individual flow-rate change. This time can be approximated by two equations [18, 66]:

$$t_s \cong \frac{380\phi\mu c_t A}{k} \tag{2.50}$$

$$t_s \cong \frac{0.04\phi\mu c_t r_e^2}{k} \tag{2.51}$$

where t_s is the stabilization time in hours, k is the permeability in md, ϕ is the porosity as a fraction, μ is viscosity in cp, c_t is total compressibility in psi^{-1}, A is area in ft^2, and r_e is the external radius in feet which should be based on the distance to the farthest drainage boundary for the well. For large systems or reservoirs with low permeability, very long stabilization times may be required.

Equation 2.49 assumes that productivity index does not change with flow rate of time, and in some wells the flow rate will remain proportional to the pressure drawdown over a wide range of flow rates. However, in many wells, the direct relationship is not linear at high flow rates as shown in Figure 2.68. The causes for the deviation in the straight-line behavior can include insufficient producing times at each rate, an increase in gas saturation near the wellbore caused by the pressure drop in that region, a decrease in permeability of oil due to the presence of gas, a reduction in permeability due to changes in formation compressibility, an increase in oil viscosity with pressure drop below the bubble point, and possible turbulence at high rates of flow.

A plot of oil production rate versus bottomhole pressure, termed the inflow performance relationship (IPR), was proposed as a method of analysis of flowing and gas-lift wells [44]. Vogel [45] calculated dimensionless IPR curves for solution gas reservoirs that covered a wide range of oil PVT and relative permeability characteristics. From computer simulations, Vogel [45] showed that any solution gas drive reservoir operating below the bubble point could be represented as shown by Figure 2.69 or by the following relationship:

$$\frac{q_o}{(q_o)_{max}} = 1 - 0.2\frac{p_{wf}}{\bar{p}} - 0.8\left(\frac{p_{wf}}{\bar{p}}\right)^2 \tag{2.52}$$

where q_o is the oil flow rate in STB/D occurring at bottomhole pressure p_{wf}, $(q_o)_{max}$ is the maximum oil flow rate in STB/D, p_{wf} is the flowing bottomhole pressure, and $\bar{p}$ is the average reservoir pressure. From the well pressure and average reservoir pressure, the ratio of producing rate to maximum oil rate can be obtained; then from the measured production rate, the maximum oil production can be calculated.

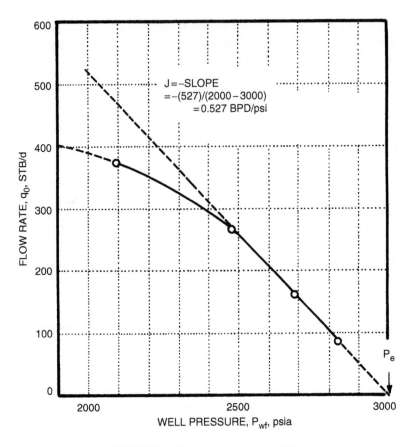

FIGURE 2.68   Productivity index [18].

Vogel's method handles the problem of a single well test when the permeability near a wellbore is the same as the permeability throughout the reservoir. When a zone of altered permeability exists near the wellbore, the degree of damage (or improvement) is expressed in terms of a "skin effect" or "skin factor." A modification to Vogel's IPR curves has been proposed by Standing [45] for situations when a skin effect is present (see Figure 2.70). In this figure, Standing has provided a series of IPR curves for flow efficiencies between 0.5 and 1.5, where flow efficiency (FE) is defined as:

$$FE = \frac{\bar{p} - p_{wf} - \Delta p_s}{\bar{p} - p_{wf}} \tag{2.53}$$

where $\bar{p}$ is the average reservoir pressure, P_{wf} is the flowing bottomhole pressure, and Δp_s is the pressure drop in the skin region. Thus, the Vogel curve is for a flow efficiency of 1.0.

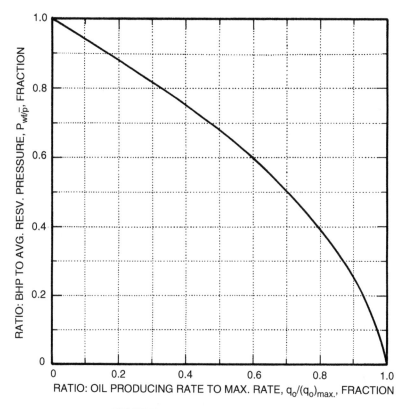

FIGURE 2.69   Vogel IPR curve [45].

For reservoir systems operating below the bubble point when fluid properties and relative permeabilities vary with distance from the wellbore,
Fetkovich [47] has proposed an empirical equation which combines single-
phase and two-phase flow:

$$q_o = J'_o(p_b^2 - p_{wf}^2)^n + J_o(p_e - p_b) \qquad (2.54)$$

where p_b is the bubble-point pressure in psia, $\bar{p}$ may be substituted for p_e,
J'_o is a form of productivity index and n is an exponent; both J'_o and n are
determined from individual well multirate and pressure tests, or isochronal
tests. For cases where the data required by the Fetkovich procedure are not
available, a method for shifting the axes of the Vogel plot has been proposed
[48]. In this latter method, only one set of production test data (rate and bottomhole flowing pressure) together with the shut-in bottomhole pressure
(or average reservoir pressure) and bubble-point pressure are required to
construct a reliable IPR.

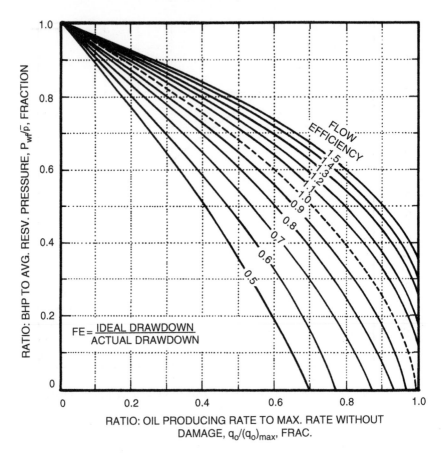

FIGURE 2.70 Standing's modification to IPR curves [46].

References

[1] Timmerman, E. H., *Practical Reservoir Engineering*, PennWell Books, Tulsa (1982).
[2] Keelan, D., "Coring: Part 1—Why It's Done," *World Oil* (March 1985), pp. 83–90.
[3] Part, A., "Coring: Part 2—Core Barrel Types and Uses," *World Oil* (April 1985), pp. 83–90.
[4] Park, A., "Coring: Part 3—Planning the Job," *World Oil* (May 1985), pp. 79–86.
[5] Park, A., "Coring: Part 4—Bit Considerations," *World Oil* (June 1985), pp. 149–154.
[6] Park, A., "Coring: Part 5—Avoiding Potential Problems," *World Oil* (July 1985), pp. 93–98.
[7] Toney, J. B., and Speights, J. L., "Coring: Part 6— Sidewall Operations," *World Oil* (Aug. 1, 1985), pp. 29–36.
[8] Kraft, M., and Keelan, D., "Coring: Part 7—Analytical Aspects of Sidewall Coring," *World Oil* (Sept. 1985), pp. 77–90.
[9] Keelan, D., "Coring: Part 8—Plug and Full Diameter Analysis," *World Oil* (Nov. 1985), pp. 103–112.
[10] Kersey, D. G., "Coring: Part 9—Geological Aspects," *World Oil* (Jan. 1986), pp. 103–108.
[11] Keelan, D. K., "Core Analysis for Aid in Reservoir Description" *J. Pet. Tech.* (Nov. 1982), pp. 2483–2491.

[12] Anderson, G., *Coring and Core Analysis Handbook*, Pet. Pub. Co., Tulsa (1975).
[13] "API Recommended Practice for Core-Analysis Procedure," API RP 40, API Prod. Dept., Dallas (Aug. 1960).
[14] "Recommended Practice for Determining Permeability of Porous Media," API RP 27, third edition, API Prod. Dept., Dallas (Aug. 1956).
[15] Stosur, J. J., and Taber, J. J., "Critical Displacement Ratio and Its Effect on Wellbore Measurement of Residual Oil Saturation," paper SPE 5509 presented at the SPE-AIME 50th Annual Fall Meeting, Dallas, Sept. 28–Oct. 1, 1975.
[16] Jenks, L. H., et al., "Fluid Flow Within a Porous Medium Near a Diamond Core Bit," *J. Can. Pet. Tech.*, Vol. 7 (1968), pp. 172–180.
[17] Black, W. M., "A Review of Drill-Stem Testing Techniques and Analysis," *J. Pet. Tech.* (June 1956), pp. 21–30.
[18] Slider, H. C., *Worldwide Practical Petroleum Reservoir Engineering Methods*, PenWell Pub. Co., Tulsa (1983).
[19] Timur, A., "An Investigation of Permeability, Porosity, and Residual Water Saturation Relationships for Sandstone Reservoirs," *The Log Analyst* (July–Aug. 1968).
[20] *Log Interpretation Charts*, Schlumberger Well Services (1991).
[21] Hilchie, D. W., *Applied Openhole Log Interpretation*, second edition, Douglas W. Hilchie, Inc., Golden, CO (1982).
[22] Doll, H. G., "The SP Log: Theoretical Analysis and Principles of Interpretation," *Trans.*, AIME, Vol. 179 (1948), pp. 146–185.
[23] Hilchie, D. W., *Old Electrical Log Interpretation*, Douglas W. Hilchie, Inc., Golden, CO (1979).
[24] Frank, R. W., *Prospecting with Old E-Logs*, Schlumberger, Houston (1986).
[25] Poupon, A., Loy, M. E., and Tixier, M. P., "A Contribution to Electrical Log Interpretations in Shaly Sands," *J. Pet. Tech.* (June 1954); Trans., AIME, pp. 27–34.
[26] Suau, J., et al., "The Dual Laterolog-R_{xo} tool," paper SPE 4018 presented at the 47th Annual Meeting, San Antonio (1972).
[27] Asquith, G., and Gibson, C., "Basic Well Log Analysis for Geologists," American Association of Petroleum Geologists, Methods in Exploration Series, Tulsa.
[28] Serra, O., Baldwin, J., and Quirein, J., "Theory Interpretation and Practical Applications of Natural Gamma Spectroscopy," *Trans.*, SPWLA (1980).
[29] Wyllie, M. R. J., Gregory, A. R., and Gardner, L. W., "Elastic Wave Velocities in Heterogeneous and Porous Media," *Geophysics*, Vol. 21, No. 1 (Jan. 1956), pp. 41–70.
[30] Merkel, R. H., "Well Log Formation Evaluation," American Association of Petroleum Geologists, Continuing Education Course-Note Series #4.
[31] Raymer, L. L., Hunt, E. R., and Gardner, J. S., "An Imporved Sonic Transit Time-To-Porosity Transform," *Trans.*, SPWLA (1980).
[32] Wahl, J. S., et al., "The Dual Spacing Formation Density Log," *J. Pet. Tech.* (Dec. 1964), pp. 1411–1416.
[33] Desbrandes, R., *Encyclopedia of Well Logging*, Gulf Publishing Co., Houston (1985).
[34] Tittman, J., "Geophysical Well Logging," *Methods of Experimental Physics*, Academic Press, Vol. 24 (1986).
[35] *Electromagnetic Propagation Tool*, Schlumberger, Ltd., Houston (1984).
[36] Bateman, R. M., "Open-Hole Log Analysis and Formation Evaluation," Intl. Human Resources Development Corp. (1985).
[37] Bateman, R. M., "Log Quality Control," International Human Resources Development Corp. (1985).
[38] Bateman, R. M., "Cased-Hole Log Analysis and Reservoir Performance Monitoring," Intl. Human Resources Development Corp. (1985).
[39] Schultz, W. E., and Smith, H. D., Jr., "Laboratory and Field Evaluation of a Carbon/Oxygen (C/O) Well Logging System," *J. Pet. Tech.* (Oct. 1974), pp. 1103–1110.

[40] Lock, G. A., and Hoyer, W. A., "Carbon-Oxygen (C/O) Log: Use and Interpretation," *J. Pet. Tech.* (Sept. 1974), pp. 1044–1054.

[41] *Cased Hole Applications*, Schlumberger, Ltd., New York (1975).

[42] *Production Log Interpretation*, Schlumberger, Ltd., New York (1973).

[43] George, C. J., and Stiles, L. H., "Improved Techniques for Evaluating Carbonate Waterfloods in West Texas," *J. Pet. Tech.* (Nov. 1978), pp. 1547–1554.

[44] Gilbert, W. E., "Flowing and Gas-lift Well Performance," *Drill. & Prod. Prac.*, API (1955), pp. 126–157.

[45] Vogel, J. V., "Inflow Performance Relationships for Solution-Gas Drive Wells," *Trans.*, AIME (1968), pp. 83–92.

[46] Standing, M. B., "Inflow Performance Relationships for Damaged Wells Producing by Solution-Gas Drive," *J. Pet. Tech.* (Nov. 1970), pp. 1399–1400.

[47] Fetkovich, M. J., "The Isochronal Testing of Oil Wells," paper SPE 4529 presented at the SPE 48th Annual Fall Meeting, Las Vegas, Sept. 30–Oct. 3, 1973.

[48] Patton, L. D., and Goland, M., "Generalized IPR Curves for Predicting Well Behavior," *Pet. Eng. International* (Sept. 1980), pp. 92–102.

[49] Amyx, J. W., Bass, D. M., Jr., and Whiting, R. L., *Petroleum Reservoir Engineering*, McGraw-Hill Book Co., Inc., New York (1960).

[50] Smith, H. I., "Estimating Flow Efficiency From Afterflow-Distorted Pressure Buildup Data," *J. Pet. Tech.* (June 1974), pp. 696–697.

[51] *Petroleum Production Handbook*, T. C. Frick (Ed.), Vol. II, Reservoir Engineering, SPE, Dallas (1962).

[52] *Log Interpretation–Principles and Applications*, Schlumberger Educational Services, Houston (1972).

[53] Kazemi, H., "Determination of Waterflood Residual Oil Saturation from Routine Core Analysis," *J. Pet. Tech.* (1977), pp. 31–32.

CHAPTER

3

Mechanisms & Recovery of Hydrocarbons by Natural Means

3.1 PETROLEUM RESERVOIR DEFINITIONS [9]

Accumulations of oil and gas occur in underground traps that are formed by structural and/or stratigraphic features. A reservoir is the portion of the trap that contains the oil and/or gas in a hydraulically connected system. Many reservoirs are hydraulically connected to water-bearing rocks or aquifers that provide a source of natural energy to aid in hydrocarbon recovery. Oil and gas may be recovered by: fluid expansion, fluid displacement, gravitational drainage, and/or capillary expulsion. In the case of a reservoir with no aquifer (which is referred to as a volumetric reservoir), hydrocarbon recovery occurs primarily by fluid expansion, which, in the case of oil, may be aided by gravity drainage. If there is water influx or encroachment from the aquifer, recovery occurs mainly by the fluid displacement mechanism which may be aided by gravity drainage or capillary expulsion. In many instances, recovery of hydrocarbon occurs by more than one mechanism.

At initial conditions, hydrocarbon fluids in a reservoir may exist as a single phase or as two phases. The single phase may be a gas phase or a liquid phase in which all of the gas present is dissolved in the oil. When there are hydrocarbons vaporized in the gas phase which are recoverable as liquids at the surface, the reservoir is called gas-condensate, and the produced liquids are referred to as condensates or distillates. For two-phase accumulations, the vapor phase is termed the gas cap and the underlying liquid phase is called the oil zone. In the two-phase ease, recovery of hydrocarbons includes the free gas in the gas cap, gas evolving from the oil (dissolved gas), recoverable liquid from the gas cap, and crude oil from the oil zone. If an aquifer or region of high water saturation is present, a transition zone can exist in which the water saturation can vary as a function of vertical depth and formation permeability. Water that exists in the oil- or gas-bearing portion of the reservoir above the transition zone is called connate or interstitial water. All of these factors are important in the evaluation of the hydrocarbon reserves and recovery efficiency.

3.2 NATURAL GAS RESERVOIRS [9]

For reservoirs where the fluid at all pressures in the reservoir or on the surface is a single gaseous phase, estimates of reserves and recoveries are relatively simple. However, many gas reservoirs produce some hydrocarbon liquid or condensate. In the latter case, recovery calculations for the single-phase case can be modified to include the condensate if the reservoir fluid remains in a single phase at all pressures encountered. However, if the hydrocarbon liquid phase develops in the reservoir, additional methods are necessary to handle these retrograde, gas-condensate reservoirs.

3.3 PRIMARY RECOVERY OF CRUDE OIL

Initial crude oil production often takes place by the expansion of fluids which were trapped under pressure in the rock. The expanding fluids may be gas evolving from the oil, an expanding gas cap, a bottom- or edge-water drive, or a combination of these mechanisms. After the initial pressure in the reservoir falls to a low value, the oil no longer flows to the wellbore, and pumps are installed to lift the crude oil to the surface. This mode of oil production is referred to as primary production. Recovery of oil associated with natural reservoir energy varies with producing mechanisms that are broadly classified as: solution-gas or depletion drive, gas cap drive, natural water drive, gravity drainage, and compaction drive. In some reservoirs, production can be attributed mainly to one of the mechanisms; in other cases, production may result from more than one mechanism, and this is referred to as a combination drive.

3.3.1 Statistical Analysis of Primary Oil Recovery

Most of the producing mechanisms are sensitive to the rate of oil production; only the solution gas drive mechanism is truly rate-insensitive [10]. Primary recoveries are usually reported [10] to be less than 25% of the original oil in place by solution gas drive, 30% to 50% of OOIP for water drive, and can exceed 75% of OOIP for gravity drainage in thick reservoirs with high vertical permeabilities. For water drive reservoirs, primary recovery efficiency can be low if the initial water saturation is more than 50%, if permeability is low, or if the reservoir is oil-wet [10]. From a statistical analysis [1], primary recovery from carbonate reservoirs tends to be lower than for sandstones (see Table 3.1 for recoveries by different drive mechanisms). Since primary recoveries tend to be lower for solution gas drive, these reservoirs are usually better candidates for waterflooding and will represent the bulk of the prospective candidates for enhanced oil recovery.

In the United States, much of the primary production involves solution gas reservoirs. Thus, this mechanism will be emphasized in this chapter, but non-U.S. production may involve other mechanisms. The differences in recovery mechanisms are important if an engineer is to avoid misapplication of methods; this subject has been addressed in Reference 11.

TABLE 3.1   Primary Recovery Efficiencies

Production mechanism	Lithology	State	Average primary recovery efficiency (% OOIP)
Solution gas drive	Sandstones	California	22
		Louisiana	27
		Oklahoma	19
		Texas 7C, 8, 10	15
		Texas 1-7B, 9	31
		West Virginia	21
		Wyoming	25
Solution gas drive	Carbonates	All	18
Natural water drive	Sandstones	California	36
		Louisiana	60
		Texas	54
		Wyoming	36
Natural water drive	Carbonates	All	44

From Reference 1.

3.3.2 Empirical Estimates of Primary Oil Recovery

Several attempts have been made to correlate primary oil recovery with reservoir parameters [1–4]. Based on field data [3] from water-drive reservoirs, a statistical study [4] yielded the following empirical relationship for primary oil recovery:

$$N_p = (0.2719 \log k + 0.25569 S_w + 0.1355 \log \mu_o$$
$$- 15,380 \phi - 0.00035 h + 0.11403)$$
$$\times \left[7,758 \, Ah\phi \, \frac{(1 - S_w)}{B_{oi}} \right] \tag{3.1}$$

where N_p is oil production in STB, k is permeability in md, S_w is fractional water saturation, μ_o is oil viscosity in cp, ϕ is fractional porosity, h is pay thickness in ft, A is a real extent in ft^2, and B_{oi} is the initial formation volume factor of oil in reservoir barrels per STB. Based on the first API study [2], correlations were developed for recoverable oil. For solution gas drive reservoirs, the recoverable oil (RO) in stock tank barrels per net acre-ft was:

$$RO = 3,244 \left[\frac{\phi(1 - S_w)}{B_{ob}} \right]^{1.1611} \left[\frac{k}{\mu_{ob}} \right]^{0.0979}$$
$$\times [S_w]^{0.3722} \left[\frac{P_b}{P_a} \right]^{0.174} \tag{3.2}$$

For water drive reservoirs, the correlation was:

$$RO = 4,259 \left[\frac{\phi(1 - S_w)}{B_{oi}} \right]^{1.0422} \left[\frac{k\mu_{wi}}{\mu_{oi}} \right]^{0.0770}$$
$$\times [S_w]^{-0.1903} \left[\frac{P_i}{P_a} \right]^{-0.2159} \tag{3.3}$$

In the second API study [1], analysis of 116 solution gas drive reservoirs gave the following equation:

$$RO = 6,533 \left[\frac{\phi(1 - S_w)}{B_{ob}} \right]^{1.312} \left[\frac{k}{\mu_{ob}} \right]^{0.0816}$$
$$\times [S_w]^{0.463} \left[\frac{P_b - P_a}{P_b} \right]^{0.249} \tag{3.4}$$

However, the second study concluded that none of the *equations developed in either study was statistically appropriate to provide a valid correlation*. Furthermore, no statistically valid correlation was found between oil recovery and

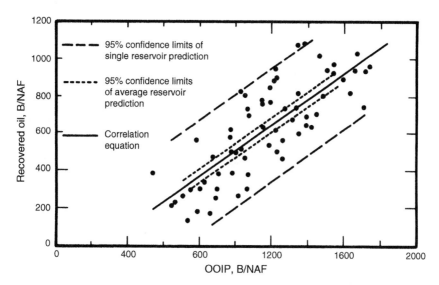

FIGURE 3.1 Correlation of primary oil recovery for water-drive reservoirs [1].

definable reservoir parameters. The second study found that when reservoirs were separated by lithology, geographical province, and producing mechanism, the only reasonable correlations that could be developed were between recoverable oil and original oil in place. Even then, the correlations were of poor quality as indicated by Figure 3.1 which presents the best correlation for Texas sandstone natural-water-drive reservoirs. The average primary recovery for various groups of reservoirs at the average value of OOIP for each group is listed by production mechanism in Table 3.1 [1, 5].

In view of the lack of suitable correlations, primary oil recovery for an individual reservoir must be estimated by one of three methods: (1) material balance equations in conjunction with equations for gas-oil ratio and fluid saturations, (2) volumetric equations if residual oil saturation and oil formation volume factor at abandonment are known or estimated, and (3) decline curve analysis, if production history is available.

3.4 PRIMARY RECOVERY FACTORS IN SOLUTION-GAS-DRIVE RESERVOIRS

Primary recovery from solution-gas-drive reservoirs depends on: type of geologic structure, reservoir pressure, gas solubility, fluid gravity, fluid viscosity, relative permeabilities, presence of connate water, rate of withdrawal, and pressure drawdown. From a statistical study [6, 7] the primary recovery factors in Table 3.2 were obtained for different oil gravities and solution gas-oil ratios in sands sandstones, limestones, dolomite, and chert. Based on work of the same type in 135 reservoir systems, Wahl[8] presented a series of figures that can be used to estimate primary recovery. One of these figures, for a condition of a 2 cp reservoir oil and a 30% connate water saturation, is reproduced in Figure 3.2. To use these figures the following is

TABLE 3.2 Primary Recovery in Percent of Oil in Place for Depletion-Type Reservoirs

Oil solution GOR ft³/bbl	Oil gravity °API	Sand or sandstones			Limestone, Dolomite or Chert		
		maximum	average	minimum	maximum	average	minimum
60	15	12.8	8.6	2.6	28.0	4.0	0.6
	30	21.3	15.2	8.7	32.8	9.9	2.9
	50	34.2	24.8	16.9	39.0	18.6	8.0
200	15	13.3	8.8	3.3	27.5	4.5	0.9
	30	22.2	15.2	8.4	32.3	9.8	2.6
	50	37.4	26.4	17.6	39.8	19.3	7.4
600	15	18.0	11.3	6.0	26.6	6.9	1.9
	30	24.3	15.1	8.4	30.0	9.6	(2.5)
	50	35.6	23.0	13.8	36.1	15.1	(4.3)
1,000	15	–	–	–	–	–	–
	30	34.4	21.2	12.6	32.6	13.2	(4.0)
	50	33.7	20.2	11.6	31.8	12.0	(3.1)
2,000	15	–	–	–	–	–	–
	30	–	–	–	–	–	–
	50	40.7	24.8	15.6	32.8	(14.5)	(5.0)

From References 7.

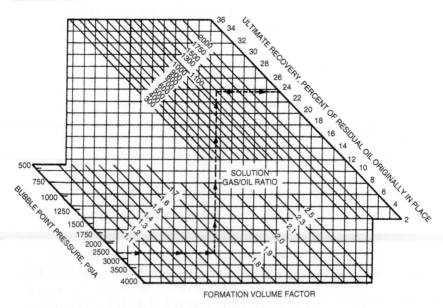

FIGURE 3.2 Estimates of primary recovery for a solution-gas-drive reservoir [8].

required: oil viscosity at reservoir conditions, interstitial water saturation, bubble-point pressure, solution gas-oil ratio at the bubble-point pressure, and formation volume factor.

References

[1] "Statistical Analysis of Crude Oil Recovery and Recovery Efficiency," API BUL D14, second edition, API Prod. Dept., Dallas (Apr. 1984).

[2] "A Statistical Study of Recovery Efficiency," API BUL D14, API Prod. Dept., Dallas (Oct. 1967).

[3] Craze, R. C., and Buckley, S. E., "A Factual Analysis of the Effect of Well Spacing on Oil Recovery," *Drill. & Prod. Prac.*, API (1945), pp. 144–159.

[4] Guthrie, R. K., and Greenberger, M. H., "The Use of Multiple-Correlation Analyses for Interpreting Petroleum-engineering Data," *Drill. & Prod. Prac.*, API (1955), pp. 130–137.

[5] Doscher, T. M., "Statistical Analysis Shows Crude-Oil Recovery," *Oil & Gas J.* (Oct. 29, 1984), pp. 61–63.

[6] Arps, J. J., and Roberts, T. G., "The Effect of the Relative Permeability Ratio, the Oil Gravity, and the Solution Gas-Oil Ratio on the Primary Recovery from a Depletion Type Reservoir," *Trans.*, AIME, Vol. 204 (1955), pp. 120–127.

[7] Arps, J. J., "Estimation of Primary Oil Reserves," *Trans.*, AIME, Vol. 207 (1956), pp. 182–191.

[8] Wahl, W. L., Mullins, L. D., and Elfrink, E. B., "Estimation of Ultimate Recovery from Solution Gas-Drive Reservoirs," *Trans.*, AIME, Vol. 213 (1958), pp. 132–138.

[9] Craft, B. C., and Hawkins, M. F., *Applied Petroleum Reservoir Engineering*, Prentice-Hall, Inc., Englewood Cliffs (1959).

[10] Timmerman, E. H., *Practical Reservoir Engineering*, PennWell Books, Tulsa (1982).

[11] Slider, H. C., *Worldwide Practical Petroleum Reservoir Engineering Methods*, PenWell Pub. Co., Tulsa (1983).

4

Fluid Movement in Waterflooded Reservoirs

Many of the principles discussed in this chapter also apply to immiscible gas injection, primary recovery by gravity drainage, and natural bottom-water drive. However, because of the importance of waterflooding in the United States, the emphasis is placed on fluid movement in waterflooded reservoirs.

The importance of various factors that affect displacement of oil by water were discussed in chapter 1. In particular, the discussion on the effect

241

of wettability on relative permeability characteristics is important in the understanding of oil displacement during waterflooding.

Several textbooks on waterflooding are available [1–3, 44]. The source most often referred to in this chapter is the excellent SPE monograph by Craig [44]; many of the principles in this monograph are summarized in the Interstate Oil Compact Commission text [2] and in an SPE paper [4]. Text from Willhite [3] contains a more thorough and mathematical treatment of the subject.

4.1 DISPLACEMENT MECHANISMS

Under ideal conditions, water would displace oil from pores in a rock in a piston-like manner or at least in a manner representing a leaky piston. However, because of various wetting conditions, relative permeabilities of water and oil are important in determining where flow of each fluid occurs, and the manner in which oil is displaced by water. In addition, the higher viscosity of crude oil in comparison to water will contribute to nonideal displacement behavior. Several concepts will be defined in order that an understanding of displacement efficiencies can be achieved.

4.1.1 Buckley–Leverett Frontal Advance

By combining the Darcy equations for the flow of oil and water with the expression for capillary pressure, Leverett [45] provided an equation for the fractional flow of water, f_w at any point in the flow stream:

$$f_w = \frac{1 + \dfrac{kk_{ro}}{v_t\mu_o}\left(\dfrac{\partial P_c}{\partial L} - g\Delta\rho\sin\alpha_d\right)}{1 + \dfrac{\mu_w}{\mu_o}\dfrac{k_o}{k_w}} \tag{4.1}$$

where f_w = fraction of water in the flowing stream passing any point in the rock (i.e., the water cut)
k = formation permeability
k_{ro} = relative permeability to oil
k_o = effective permeability to oil
k_w = effective permeability to water
μ_o = oil viscosity
μ_w = water viscosity
v_t = total fluid velocity (i.e., q_t/A)
P_c = capillary pressure = $p_o - p_w$ = pressure in oil phase minus pressure in water phase

L= distance along direction of movement

g= acceleration due to gravity

$\Delta\rho$= water-oil density differences $= \rho_w - \rho_o$

α_d = angle of the formation dip to the horizontal.

This equation is derived in an appendix in the monograph by Craig [44]. Because relative permeabilities and capillary pressure are functions of only fluid saturation, the fractional flow of water is a function of water saturation alone. In field units, Equation 4.1 becomes [44]:

$$f_w = \frac{1 + 0.001127 \dfrac{k k_{ro}}{\mu_o} \dfrac{A}{q_t} \left(\dfrac{\partial P_c}{\partial L} - 0.433 \Delta\rho \sin \alpha_d \right)}{1 + \dfrac{\mu_w}{\mu_o} \dfrac{k_o}{k_w}} \quad (4.2)$$

where permeability is in md, viscosities are in cp, area is in sq ft, flow rate is in B/D, pressure is in psi, distance is in ft, and densities are in g/cc.

In practical usage, the capillary pressure term in Equation 4.1 is neglected [44]:

$$f_w = \frac{1 - \dfrac{k\, k_{ro}}{v\, \mu_o}(g\, \Delta\rho \sin \alpha_d)}{1 + \dfrac{\mu_w}{\mu_o} \dfrac{k_o}{k_w}} \quad (4.3)$$

and for a horizontal displacement of oil by water, the simplified form of this equation is [44]:

$$f_w = \frac{1}{1 + \dfrac{\mu_w}{\mu_o} \dfrac{k_{ro}}{k_{rw}}} \quad (4.4)$$

Examples of idealized fractional flow curves, f_w vs. S_w, are given in Figure 4.1 for strongly water-wet and strongly oil-wet conditions [44].

Based on the initial work of Leverett [45], Buckley and Leverett [46] presented equations to describe an immiscible displacement in one-dimensional flow. For incompressible displacement, the velocity of a plane of constant water saturation traveling through a linear system was given by:

$$v = \frac{q}{A\phi} \left(\frac{\partial f_w}{\partial S_w} \right) \quad (4.5)$$

where q is the flow rate in cc/sec (or ft^3/D), A is the cross-sectional area in cm^2 (or ft^2), ϕ is the fractional porosity, v is the velocity or rate of advance in cm/sec (or ft/D), and $(\partial f_w/\partial S_w)$ is the slope of the curve of f_w vs. S_w. This equation states that the rate of advance or velocity of a plane of constant water saturation is directly proportional to the derivative of the water cut

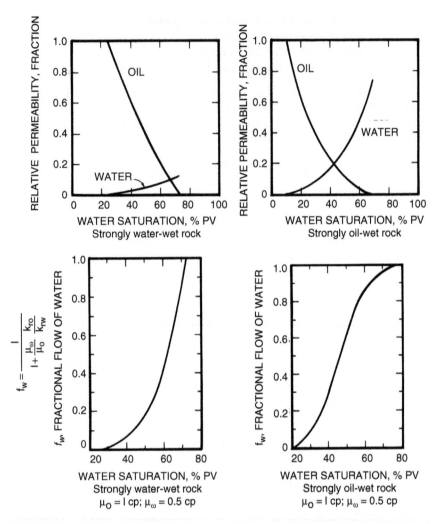

FIGURE 4.1 Effect of water saturation on relative permeabilities and fractional flow of water [44].

at that water saturation. By integrating Equation 4.5 for the total time since the start of injection, the distance that the plane of given water saturation moves can be given by:

$$L = \frac{W_i}{A\phi}\left(\frac{\partial f_w}{\partial S_w}\right) \tag{4.6}$$

where W_i is the cumulative water injected and L is the distance that a plane of given saturation has moved.

If L is the distance from injector to producer, the time of water break-through, t_{bt}, is given by:

$$t_{bt} = \frac{L}{\dfrac{q}{A\phi}\left(\dfrac{\partial f_w}{\partial S_w}\right)} \tag{4.7}$$

Equation 4.6 can be used to calculate the saturation distribution in a linear waterflood as a function of time. According to Equation 4.6, the distance moved by a given saturation in a given time interval is proportional to the slope of the fractional flow curve at the saturation of interest. If the slope of the fractional flow curve is graphically obtained at a number of saturations, the saturation distribution in the reservoir can be calculated as a function of time. The saturation distribution can then be used to predict oil recovery and required water injection on a time basis. A typical plot of df_w/dS_w vs. S_w will have a maximum as shown in Figure 4.2. However, a problem is that

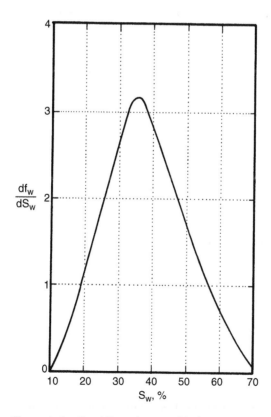

FIGURE 4.2 Change in fractional flow of water with change in water saturation [44].

equal values of the slope, df_w/dS_w, can occur at two different saturations which is not possible. To overcome this difficulty, Buckley and Leverett [46] suggested that a portion of the saturation distribution curve is imaginary, and that the real curve contains a saturation discontinuity at the front. Since the Buckley-Leverett procedure neglects capillary pressure, the flood front in a practical situation will not exist as a discontinuity, but will exist as a stabilized zone of finite length with a large saturation gradient.

4.1.2 Welge Graphical Technique

A more simplified graphical technique was proposed by Welge [47] which involves integrating the saturation distribution from the injection point to the front. The graphical interpretation of this equation is that a line drawn tangent to the fractional flow curve from the initial water saturation (S_{wi}) will have a point of tangency equal to water saturation at the front (S_{wf}). Additionally, if the tangent line is extrapolated to $f_w = 1$, the water saturation will correspond to the average water saturation in the water bank, $\bar{S}_w$. Construction of a Welge plot is shown in Figure 4.3. The tangent line should be drawn from the initial water saturation even if that saturation is greater than the irreducible water saturation.

Welge derived an equation that relates the average displacing fluid saturation to the saturation at the producing end of the system:

$$\bar{S}_w - S_{w2} = Q_i f_{o2} \tag{4.8}$$

where $\bar{S}_w$ = average water saturation, fraction of PV
  S_{w2} = water saturation at the producing end of the system, fraction of PV
  Q_i = pore volumes of cumulative injected fluid, dimensionless
  f_{o2} = fraction of oil flowing at the outflow end of the system

Equation 4.8 is important because it relates to three factors of prime importance in waterflooding [44]: (1) the average water saturation and thus the total oil recovery, (2) the cumulative injected water volume, and (3) the water cut and hence the oil cut.

Welge also related the cumulative water injected and the water saturation at the producing end:

$$Q_t = \frac{1}{\left(\dfrac{df_w}{dS_w}\right)_{S_{w2}}} \tag{4.9}$$

Thus, the reciprocal of the slope of the tangent line gives the cumulative water influx at the time of water breakthrough. When a value of Q_i and the

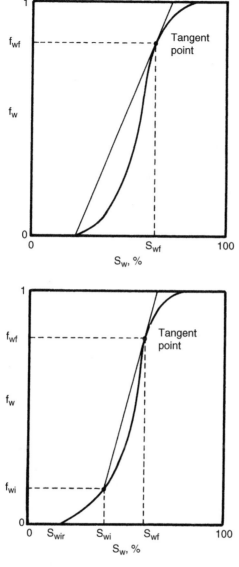

FIGURE 4.3 Welge graphical plot [47].

injection rate are known, the time to reach that stage of the flood can be computed.

For a liquid-filled, linear system, the average water saturation at breakthrough, $\bar{S}_{wbt}$, is:

$$\bar{S}_{wbt} = S_{iw} + \frac{W_i}{A\phi L} \tag{4.10}$$

where S_{iw} is the irreducible or connate water saturation. If Equation 4.6 is substituted into Equation 4.10:

$$\bar{S}_{wbt} - S_{iw} = \frac{1}{\left(\dfrac{df_w}{dS_w}\right)} = \frac{S_{wf} - S_{iw}}{f_{wf}} \qquad (4.11)$$

where S_{wf} is the water saturation at the flood front and f_{wf} is the water cut at the flood front. After breakthrough, water saturation is obtained from Equations 4.8 and 4.9 where, as mentioned earlier: (1) the tangent point, S_{w2}, represents the water saturation at the producing end of the system, (2) the value of f_w at the point of tangency is the producing water cut, (3) the saturation at which the tangent intersects $f_w = 1.0$ is the average water saturation, and (4) the inverse of the slope of the tangent line is equal to the cumulative injected fluid in pore volumes (Q_i). If connate water is mobile, appropriate corrections need to be made [44].

Oil production at breakthrough can be computed from [3]:

$$N_{pbt} = \frac{\phi AL}{B_o} (\bar{S}_{wbt} - S_{iw}) \qquad (4.12)$$

After water breakthrough, a number of saturation greater than S_{wf} are selected; the slope of the tangent line and average water saturation are determined for each value of S_w chosen. Oil production after breakthrough is then determined by observing the change in water saturation [3]:

$$\Delta N_p = \frac{\phi AL}{B_o} (\bar{S}_w - \bar{S}_{wbt}) \qquad (4.13)$$

The incremental oil production from Equation 4.13 can be added to the breakthrough production from Equation 4.12, and the resulting total production for the linear system can be listed as a function of S_w, time, or other parameters. If the pore volumes in these equations are in ft^3, divide by 5.615 to get barrels.

4.2 VISCOUS FINGERING

A problem often encountered in the displacement of oil by water is the viscosity contrast between the two fluids. The adverse mobility ratios that result promote fingering of water through the more viscous crude oil and can reduce the oil recovery efficiency. An example of viscous fingering is shown in Figure 4.4.

INJECTION
WELL

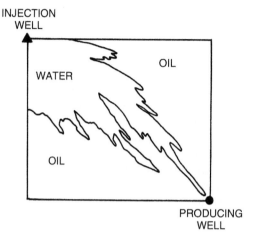

FIGURE 4.4   Viscous fingering.

PRODUCING
WELL

4.3 MOBILITY AND MOBILITY RATIO

Mobility of a fluid is defined as the ratio of the permeability of the formation to a fluid, divided by the fluid viscosity:

$$\lambda = \frac{k}{\mu} \qquad (4.14)$$

where λ = mobility, md/cp
   k = effective permeability of reservoir rock to a given fluid, md
   μ = fluid viscosity, cp

When multiple fluids are flowing through the reservoir, relative permeabilities must be used along with viscosities of the fluids. By convention, the term mobility ratio is defined as the mobility of the displacing fluid divided by the mobility of the displaced fluid. For waterfloods, this is the ratio of water to oil mobilities. Thus the mobility ratio, M, for a waterflood is:

$$M = \frac{k_{rw}/\mu_w}{k_{ro}/\mu_o} = \frac{k_{rw}\mu_o}{k_{ro}\mu_w} \qquad (4.15)$$

where k_{rw} and k_{ro}, are relative permeabilities to water and oil, respectively, μ_o is oil viscosity and μ_w is water viscosity. Prior to 1957, there was no accepted definition, and many workers defined mobility ratio as oil to water mobility; in this case, the reciprocal of mobility ratio (as now accepted) must be used. The oil mobility used in Equation 4.10 refers to the location in the oil bank ahead of the flood front. For the water mobility, there are several possibilities regarding the location at which the relative permeability should be chosen: at the flood front, at residual oil saturation where only water is flowing (end point), or at some intermediate saturation. Craig [44] found a better correlation if the water mobility was determined at the average water saturation behind the flood front at water breakthrough. Thus for the mobility ratio expression, the relative permeability of water is found at the average water saturation at water break-through as determined by the Welge graphical approach. As Craig notes, the mobility ratio of a waterflood will remain constant before breakthrough, but it will increase after water breakthrough corresponding to the increase in water saturation and relative permeability to water in the water-contacted portion of the reservoir. Unless otherwise specified, the term mobility ratio is taken to be the value prior to water breakthrough. As will be discussed later in this chapter, mobility ratio is important in determining the volume of reservoir contacted by the waterflood.

4.4 RECOVERY EFFICIENCY

Recovery efficiency is the fraction of oil in place that can be economically recovered with a given process. The efficiency of primary recovery mechanisms will vary widely from reservoir to reservoir, but the efficiencies are normally greatest with water drive, intermediate with gas cap drive, and least with solution gas drive. Results obtained with waterflooding have also varied. The waterflood recovery can range from less than the primary recovery to as much as 2.5 times the recovery obtained in some solution-gas drive reservoirs. A recent statistical analysis by the API [48] provided the average primary and secondary recovery efficiencies in Table 4.1. Generally, primary and ultimate recoveries from carbonate reservoirs tend to be lower than from sandstones. For pattern waterfloods, the average ratio of secondary to primary recovery ranges from 0.33 in California sandstones to greater than one in Texas carbonates. For edge water injection, the secondary-to primary ratio ranged from an average of 0.33 in Louisiana to 0.64 in Texas. By comparison, secondary recovery for gas injection into a gas cap averaged only 0.23 in Texas sandstones and 0.48 in California sandstones. Ultimate primary and secondary recovery performance for different drive mechanisms are given in Table 4.2. Solution-gas-drive reservoirs will generally have higher oil saturations after primary recovery, and are usually the better candidates for waterflooding.

Displacement of oil by waterflooding is controlled by fluid viscosities, oil-water relative permeabilities, nature of the reservoir rock, reservoir heterogeneity, distribution of pore sizes, fluid saturations (especially the amount of oil present), capillary pressure, and the location of the injection wells in relation to the production wells. These factors contribute to the overall process efficiency. Oil recovery efficiency (E_R) of a waterflood is

TABLE 4.1 Ultimate Recovery (Primary Plus Secondary)

	Average OOIP B/NAF	Ultimate Recovery at Average OOIP B/NAF	Recovery Efficiency			
			Primary % OOIP	Secondary % OOIP	Ultimate % OOIP	Secondary to primary Ratio
Pattern waterfloods						
California sandstones	1,311	463	26.5	8.8	35.3	0.33
Louisiana sandstones	1,194	611	36.5	14.7	51.2	0.40
Oklahoma sandstones	728	201	17.0	10.6	27.6	0.62
Texas sandstones	942	362	25.6	12.8	38.4	0.50
Wyoming sandstones	774	346	23.6	21.1	44.7	0.89
Texas carbonates	388	123	15.5	16.3	31.8	1.05
Edge water injection						
Louisiana sandstones	1,181	680	41.3	13.8	55.1	0.33
Texas sandstones	897	499	34.0	21.6	55.6	0.64
Gas injection into cap						
California sandstones	909	396	29.4	14.2	43.6	0.48
Texas sandstones	957	412	35.3	8.0	43.3	0.23

From References 48 and 49.
Recovery Efficiency: Average value of the recoverable oil divided by the average value of the original oil-in-place for the reservoirs in the classification.
OOIP: Original oil-in-place
B/NAF: Barrels per net acre-ft.

TABLE 4.2   Secondary Recovery Efficiencies

Secondary Recovery Method	Lithology	State	Primary Plus Secondary Recovery Efficiency (% OOIP)	Ratio of Secondary to Primary Recovery Efficiency
Pattern waterflood	Sandstone	California	35	0.33
		Louisiana	51	0.40
		Oklahoma	28	0.62
		Texas	38	0.50
		Wyoming	45	0.89
Pattern waterflood	Carbonates	Texas	32	1.05
Edge water injection	Sandstone	Louisiana	55	0.33
		Texas	56	0.64
Gas cap injection	Sandstone	California	44	0.48
		Texas	43	0.23

From References 48 and 49.

the product of displacement efficiency (E_D) and volumetric efficiency (E_V), both of which can be correlated with fluid mobilities:

$$E_R = E_D E_V = E_D E_P E_I \tag{4.16}$$

where E_R = overall reservoir recovery or volume of hydrocarbons recovered divided by volume of hydrocarbons in place at start of project

E_D = volume of hydrocarbons (oil or gas) displaced from individual pores or small groups of pores divided by the volume of hydrocarbons in the same pores just prior to displacement

E_P = pattern sweep efficiency (developed from areal efficiency by proper weighting for variations in net pay thickness, porosity, and hydrocarbon saturation): hydrocarbon pore space enclosed behind the injected-fluid front divided by total hydrocarbon pore space of the reservoir or project.

E_I = hydrocarbon pore space invaded (affected, contacted) by the injection fluid or heat-front divided by the hydrocarbon pore space enclosed in all layers behind the injected fluid

4.5 DISPLACEMENT SWEEP EFFICIENCY (E_D)

Factors affecting the displacement efficiency for any oil recovery process are pore geometry, wettability (water-wet, oil-wet, or intermediate), distribution of fluids in the reservoir, and the history of how the saturation occurred. Results are displayed in the relative permeability curves (Figure 4.1) from which the flowing water saturation (or conversely the oil saturation) can be obtained at any total fluid saturation. As shown in Figure 4.5, displacement efficiencies decrease as oil viscosities increase [44].

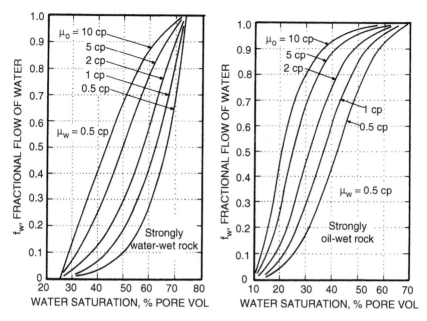

FIGURE 4.5   Effect of oil viscosity on fractional flow of water [44].

4.6 VOLUMETRIC SWEEP EFFICIENCY (E$_V$)

Whereas displacement efficiency considers a linear displacement in a unit segment (group of pores) of the reservoir, macroscopic or volumetric sweep takes into account that fluid (i.e., water) is injected at one point in a reservoir and that other fluids (i.e., oil, water) are produced from another point (Figure 4.6). Volumetric sweep efficiency, the percentage of the total reservoir contacted by the injected fluid (often called fluid conformance), is composed of areal (or pattern) efficiency and vertical sweep.

4.7 AREAL OR PATTERN SWEEP EFFICIENCY (E$_P$)

Areal sweep efficiency of an oil recovery process depends primarily on two factors: the flooding pattern and the mobilities of the fluids in the reservoir. In the early work on sweep efficiency and injectivity, Muskat and coworkers [5, 50] presented analytical solutions for direct line drive, staggered line drive, 5-spot, 7-spot, and 9-spot patterns. Experimental studies on the effect of mobility ratio for different patterns were presented by Dyes, Caudle, and Erickson [6] (5-spot and line drives); Craig, Geffen, and Morse [7], Prats et al. [8], Caudle and Witte [9], and Haberman [10]

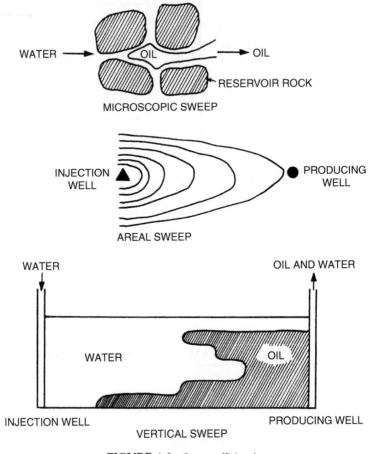

FIGURE 4.6 Sweep efficiencies.

(5-spot); and Kimbler, Caudle, and Cooper [11] (9-spot). The effect of sweep-out beyond the pattern area was studied as well [12, 13]. From a mathematical study the breakthrough sweep efficiency of the staggered line drive was presented by Prats [14]. A comparison of the areal sweep efficiency and the ratio d/a is shown in Figure 4.7 for direct and staggered line drives [14, 50], and a review of the early work was provided by Crawford [15].

Areal sweep efficiency at breakthrough for a 5-spot pattern is shown in Figure 4.8, and the effect of mobility ratio on areal sweep is shown in Figure 4.9. These figures show that areal sweep efficiency is low when mobility ratio is high (note that the data in Figure 4.9 from Dyes, Caudle and Erickson are plotted in terms of the reciprocal of mobility ratio as currently defined). Areal sweep efficiencies at breakthrough, for different patterns and a mobility ratio of one, are summarized in Table 4.3 [2, 4, 44].

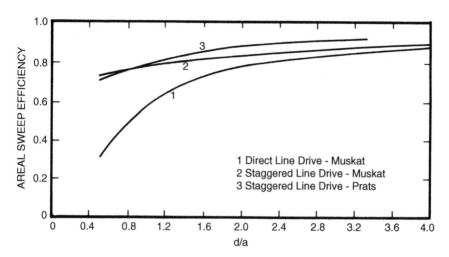

FIGURE 4.7   Areal sweep efficiencies for direct and staggered line drive patterns [14, 50].

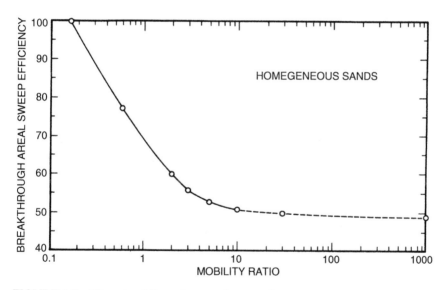

FIGURE 4.8   Effect of mobility ratio on areal sweep efficiency at breakthrough for a 5-spot pattern [44].

Areal sweep efficiency is more important for considering rate vs. time behavior of a waterflood rather than ultimate recovery because, at the economic limit, most of the interval flooded has either had enough water throughout to provide 100% areal sweep or the water bank has not yet reached the producing well so that no correction is needed for areal sweep [44].

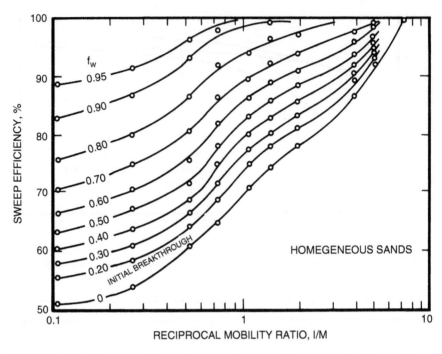

FIGURE 4.9 Effect of mobility ratio on areal sweep efficiency after breakthrough for a 5-spot pattern [6].

TABLE 4.3   Areal Sweep Efficiency at Breakthrough (M = 1)

Type of Pattern	Ep, % [4]	Ep, % [2]*
Direct line drive (d/a = 1)	57.0	57
Direct line drive (d/a = 1.5)	70.6	—
Staggered line drive	80.0	75 (d/a = 1)
5-spot	72.3	68–72
7-spot	74.0	74–82
9-spot diagonal/directional rate		
0.5	—	49
1	—	54
5	—	69
10	—	78

From Reference 44.
*Based on summary of data presented by Craig.

When waterflooding calculations are performed, especially with computers or programmable calculators, the use of equations with adjustable coefficients are very useful. Recently, Fassihi [16] provided correlations for the calculation of areal and vertical sweep efficiencies. For these correlations of areal sweep, the data of Dyes, Caudle, and Erickson [6] were curve-fitted

TABLE 4.4   Coefficients in Areal Sweep Efficiency Correlations

Coefficient	5-spot	Direct Line Drive	Staggered Line Drive
a_1	−0.2062	−0.3014	−0.2077
a_2	−0.0712	−0.1568	−0.1059
a_3	−0.511	−0.9402	−0.3526
a_4	0.3048	0.3714	0.2608
a_5	0.123	−0.0865	0.2444
a_6	0.4394	0.8805	0.3158

From Reference 16.

and the resulting equation was:

$$\frac{1-E_P}{E_P} = [a_1 \ln(M+a_2) + a_3]f_w + a_4 \ln(M+a_5) + a_6 \tag{4.17}$$

where E_P is the areal sweep efficiency which is the fraction of the pattern area contacted by water, M is the mobility ratio, and the coefficients are as listed in Table 4.4 for the 5-spot, direct line drive, and staggered line drive. These coefficients are valid both before and after breakthrough, and apply to mobility ratios between zero and ten, which is within the range observed in many waterfloods. For the 5-spot pattern, these values of E_p are generally higher than in later experiments, and a correction has been suggested by Claridge [17] that should be multiplied by the E_P from the Dyes et al. data:

$$E_v = \frac{E_P/V_d}{\{M^{0.5} - [(M-1)(1-E_P/V_d)]^{0.5}\}^2} \tag{4.18}$$

where E_V is the volumetric sweep efficiency in a linear displacement, V_d is the displaceable pore volumes injected, and the other terms are as already defined.

4.8   VERTICAL OR INVASION SWEEP EFFICIENCY (E_I)

For well-ordered sandstone reservoirs, the permeability measured parallel to the bedding planes of stratified rocks is generally larger than the vertical permeability. For carbonate reservoirs, permeability (and porosity) may have developed after the deposition and consolidation of the formation; thus the concept of a stratified reservoir may not be valid. However, in stratified rocks, vertical sweep efficiency takes into account the inherent vertical permeability variations in the reservoir. Vertical sweep efficiency of a waterflood depends primarily upon the vertical distribution of permeabilities within the reservoir, on the mobility of fluids involved, and on the density differences between flowing fluids. As a result of nonuniformity of permeabilities in the vertical direction, fluid injected into an oil-bearing formation will seek the paths of least resistance and will move through the reservoir as an irregular front. Consequently, the injected fluid will travel more rapidly in the more permeable zones and will travel less rapidly in the tighter zones. With continued injection, and displacement of some of the resident fluids, the saturation of the injected fluid will become greater in the more permeable areas than in the low-permeability strata. This can cause early breakthrough of injected fluid into the producing wells before the bulk of the reservoir has been contacted. In addition, as the saturation of the injected fluid increases in the highly permeable zones, the relative permeability to that fluid also increases. All of these effects can lead to channeling of the injected fluid, which is aggravated by the unfavorable viscosity ratio common in waterflooding. In many cases, permeability stratification has a dominant effect on behavior of the waterflood.

4.9 PERMEABILITY VARIATION

Two methods of quantitatively defining the variation in vertical permeabilities in reservoirs are commonly used. The extent of permeability stratification is sometimes described with the Lorenz coefficient [18] and is often described with the Dykstra–Parsons [19] coefficient of permeability variation.

4.9.1 Lorenz Coefficient

Schmalz and Rahme [18] suggested arranging the vertical distribution of permeabilities from highest to lowest, and plotting the fraction of total flow capacity (kh) versus the fraction of total volume (hφ). To obtain the Lorenz coefficient (see Figure 4.10), the area ABCA is divided by the area ADCA. Values of the Lorenz coefficient can range from zero for a uniform reservoir to a theoretical maximum value of one. However, the Lorenz coefficient is not a unique measure of stratification, and several different permeability distributions can give the same Lorenz coefficient [44].

4.9.2 Dykstra–Parsons Coefficient of Permeability Variation

The coefficient of permeability variation described by Dykstra and Parsons [19] is also referred to as the permeability variation or permeability variance. This method assumes that vertical permeabilities in a reservoir will have a log-normal distribution. The procedure outlined by Dykstra and Parsons was to: (1) divide permeabilities (usually from core analysis) so that all samples are of equal thickness (often 1 ft), (2) arrange the permeabilities in descending order from highest to lowest, (3) calculate for each sample the percent of samples that have a higher permeability (see example in Table 4.5), (4) plot the data from Step 3 on log-probability paper (see Figure 4.11) (5) draw the best straight line through data (with less emphasis

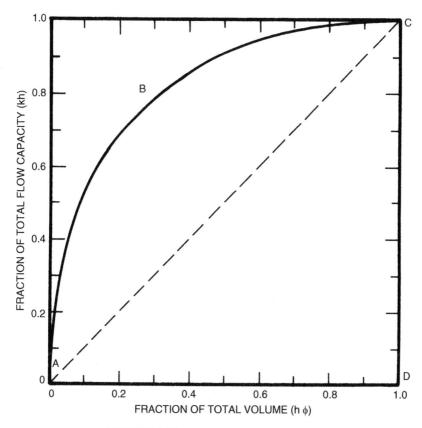

FIGURE 4.10 Lorenz coefficient plot [44].

on points at the extremities, if necessary), (6) determine the permeability at 84.1% probability ($k_{84.1}$) and the mean permeability at 50% probability (k_{50}), and (7) compute the permeability variation, V:

$$V = \frac{k_{50} - k_{84.1}}{k_{50}} \qquad (4.19)$$

As with the Lorenz coefficient, the possible values of the Dykstra–Parsons permeability variation range from zero for a uniform reservoir to a maximum value of 1. In some cases, there may be a direct relation between the Lorenz and Dykstra–Parsons coefficients [20], but in many instances a direct relationship with field data will not be observed. Often, insufficient data are available to provide enough samples for adequate analysis, and in some cases, the data may not provide a log-normal distribution.

Increasing values of permeability variation indicate increasing degrees of vertical heterogeneity in a reservoir. Permeability variations often range from about 0.5 to 0.8; lower numbers may be observed for relatively uniform

TABLE 4.5   Data for Permeability Variation plot

Permeability (md)	Percent of Samples with Greater than Stated Permeability
950	0
860	5
640	10
380	15
340	20
280	25
210	30
160	35
135	40
130	45
110	50
78	55
65	60
63	65
54	70
40	75
27	80
21	85
20	90
15	95

reservoirs, and higher numbers may be calculated for very nonuniform reservoirs. Using the data from Dykstra and Parsons, Johnson [21] provided a graphical technique to estimate recovery during an immiscible displacement. One of Johnson's plots is reproduced in Figure 4.12 for a producing water-oil ratio (WOR) of 100 which could represent the economic limit for many waterfloods. Lines of constant recovery are given as functions of permeability variation and mobilitiy. Johnson also provided plots for $WOR = 1, WOR = 5$, and $WOR = 25$. At any WOR, and increase in vertical permeability variation yielded a lower recovery. As will be discussed later under prediction methods, the Dykstra–Parsons fractional recovery, R, as a percent of oil in place, must be multipled by the areal sweep efficiency, Ep, to obtain an estimate of the oil recovered.

As mentioned earlier, correlations for calculating vertical and areal sweep efficiencies were recently provided by Fassihi [16]. The correlating parameter, Y, for vertical coverage, C, is:

$$Y = a_1 C^{a_2} (1 - C)^{a_3}$$ (4.20)

where $a_1 = 3.334088568$
$a_2 = 0.7737348199$
$a_3 = -1.225859406$

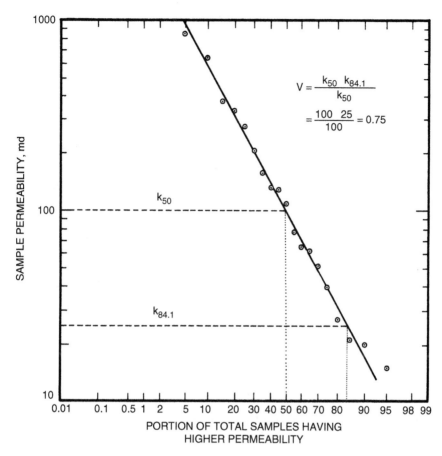

FIGURE 4.11 Dykstra–Parsons plot of permeability variation [19, 44].

where the equation for Y was given by deSouza and Brigham [22] in terms of water-oil ratio (WOR), mobility ratio (M), and permeability variation (V):

$$Y = \frac{(WOR + 0.4)(18.948 - 2.499\,V)}{(M + 1.137 - 0.8094\,V)^{f(V)}} \quad (4.21)$$

where $f(V) = -0.6891 + 0.9735\,V + 1.6453\,V^2$.

These equations are valid for mobility ratios ranging from 0 to 10 and for permeability variations ranging from 0.3 to 0.8.

Based on calculations of WOR vs. oil recovery for a 5-spot pattern, Craig [23] found that there was a minimum number of equal thickness layers required to obtain the same performance as with 100 layers. Table 4.6 shows the effect of permeability variation and mobility ratio on the minimum number of layers for WORs above 10. Craig [23] presented similar tables for lower WORs.

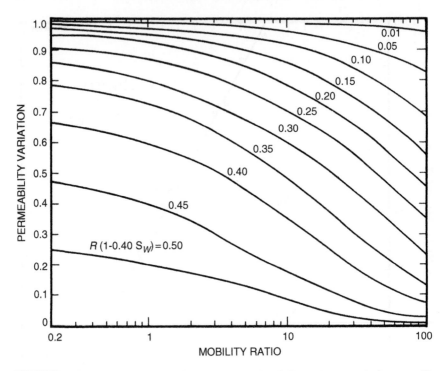

FIGURE 4.12   Effect of permeability variation and mobility ratio on vertical sweep efficiency at WOR = 100 [21].

TABLE 4.6   Minimum Number of Equal-Thickness Layers Required to Obtain Performance of a 100-Layer, 5-Spot Waterflood at Producing WOR's Above 10

Mobility Ratio	Permeability Variation							
	0.1	0.2	0.3	0.4	0.5	0.6	0.7	0.8
0.05	1	1	1	2	4	5	10	20
0.1	1	1	1	2	5	5	10	20
0.2	1	1	2	3	5	5	10	20
0.5	1	1	2	3	5	5	10	20
1.0	1	1	2	3	5	10	10	50
2.0	1	2	3	4	10	10	20	100
5.0	1	3	4	5	10	100	100	100

From Reference 23.

An analytical extension [24] of the Dykstra–Parsons method allows calculations of total flow rates and flow rates in each layer for both a constant injection rate and for a constant pressure drop. The ability to calculate cumulative injection into a layer allows the incorporation of sweep efficiency of each layer as a function of mobility ratio and displaceable pore volumes injected for the pattern used in the waterflood.

4.9.3 Crossflow

In the usual cases where there is vertical communication between the different layers of varying permeabilities, the effect of vertical crossflow must be considered [25, 26]. Goddin et al. [26] performed a numerical simulation in a 2-D, 2-layer, water-wet system. For mobility ratios ranging from 0.21 to 0.95, oil recovery with crossflow was between that computed for a uniform reservoir and that for a layered reservoir with no crossflow. Goddin et al. [26] defined a crossflow index, which is a measure of the extent the performance varies from that of a uniform permeability system:

$$\text{crossflow index} = \frac{N_{pcf} - N_{pncf}}{N_{pu} - N_{pncf}} \tag{4.22}$$

where N_{pu} = oil recovery from uniform system with the
   average permeability
N_{pcf} = oil recovery from layered system with crossflow
N_{pncf} = oil recovery from stratified system with no crossflow

Of the variables investigated, mobility ratio and the permeability ratio of the two layers had the largest effect on crossflow (see Figures 4.13 and 4.14, respectively). Crossflow was more pronounced at lower mobility ratios or at high ratios of layer permeabilities. The crossflow index of one means that the performance of the layered system with crossflow is identical to the performance of the system with uniform permeability.

Still at issue is the relative importance of mobility ratio and gravity in waterflooding stratified reservoirs [27–31]. For wetting conditions that are not strongly water-wet, additional complications will arise.

4.9.4 Estimates of Volumetric Sweep Efficiency

Volumetric sweep efficiency ranges from about 0.1 for very heterogeneous reservoirs to greater than 0.7 for homogeneous reservoirs with good flooding characteristics [3]. For a liquid-filled, 5-spot pattern, Craig [23] found that the volumetric sweep efficiency (E_V) at breakthrough decreases sharply as the permeability variation increases (see Figure 4.15). Similar trends were observed for initial gas saturations of 10% and 20%. These data indicated that the major effect of mobility ratio on E_V at breakthrough occurs for mobility ratios ranging from 0.1 to 10.

More recent simulations [32] of 5-spot patterns with a streamtube model yielded the volumetric sweep efficiencies shown in Figures 4.16 and 4.17 for WORs of 25 and 50, respectively. Mobility ratios of 0.1, 1, 10, 30, and 100 were used. The permeabilities in the 100-layer model were assumed to have a log-normal distribution, and pseudo-relative permeability expressions were used. In a companion paper [33] the streamtube model (no crossflow) was compared to the Dykstra–Parsons method (no crossflow) and with a

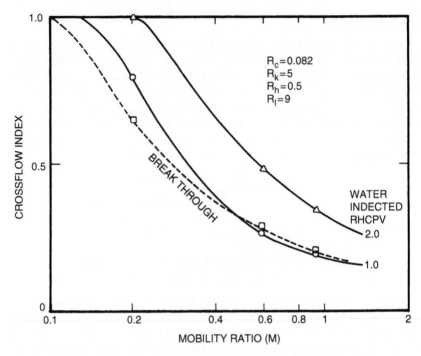

FIGURE 4.13   Influence of mobility ratio on vertical crossflow [26].

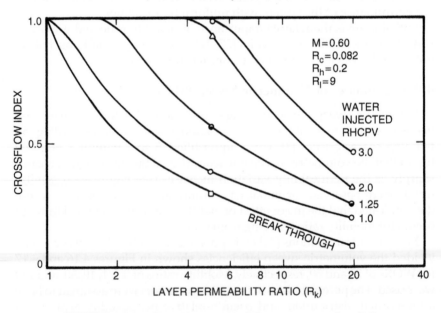

FIGURE 4.14   Effect of layer permeability ratio on crossflow [26].

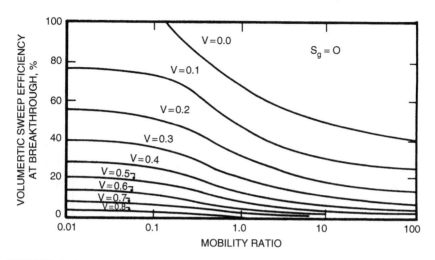

FIGURE 4.15 Effect of permeability variation and mobility ratio on volumetric sweep at breakthrough [23].

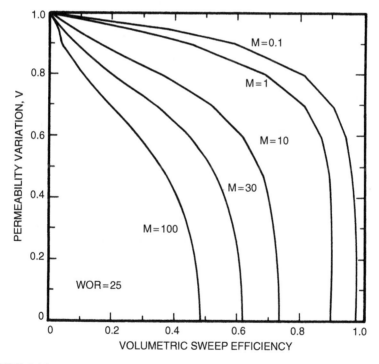

FIGURE 4.16 Effect of permeability variation and mobility ratio on volumetric sweep at breakthrough [32].

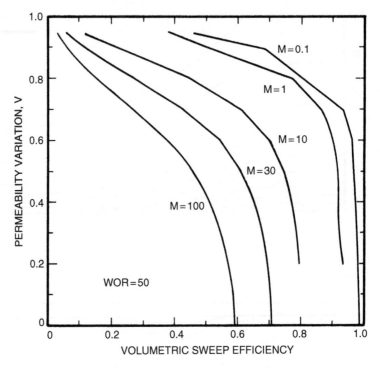

FIGURE 4.17   Effect of permeability variation and mobility ratio on volumetric sweep at WOR = 50 [32].

model having the assumption of equal pressure gradient in each layer (with crossflow). The streamtube model was more closely described by the model with vertical communication for unfavorable (high) mobility ratios and by the Dykstra–Parsons model for favorable (low) mobility ratios.

4.10   ESTIMATION OF WATERFLOOD RECOVERY BY MATERIAL BALANCE

Oil recovered by waterflooding, N_{pw}, in STB, can be estimated from [3]:

$$N_{pw} = \frac{7{,}758\,Ah\phi\left[S_{op} - E_V S_{or} - (1 - E_v)S_{oi}\right]}{B_o} \tag{4.23}$$

where 7,758 is the number of barrels per acre-ft, A is areal extent of the reservoir in acres, h is reservoir thickness in ft, ϕ is the fractional porosity, S_{op} is the oil saturation at the start of waterflooding, S_{or} is the waterflood residual oil saturation, S_{oi} is the initial oil saturation, B_o is the oil formation volume factor, and E_V is the volumetric sweep or fraction of the reservoir volume swept by the injected water when the economic limit has been reached. In terms of the original oil in place, the waterflood recovery is [3]:

$$N_{pw} = (N - N_p) - N \frac{B_{oi}}{B_o} \left[1 + E_V \left(\frac{S_{or}}{S_{oi}} - 1 \right) \right] \qquad (4.24)$$

where N_{pw} = oil potentially recoverable by waterflooding, STB
N = initial oil in place, STB
N_p = oil produced during primary operations, STB
B_{oi} = initial FVF

These equations can be altered to include a residual gas saturation, if present. The volumetric sweep efficiency can be estimated from one of the correlations given previously or can be obtained from an analogy from similar water-flood projects.

4.11 PREDICTION METHODS

An extensive survey on prediction of waterflood performance was provided by Craig [44]. Of the methods reviewed, three appeared most promising: (1) the Higgins-Leighton streamtube model [34], (2) the Craig, Geffen, and Morse model [7], and (3) the Prats et al. method [8]. Discussion of the various prediction methods is beyond the scope of this text, and only two very simple methods will be presented for illustrative purposes. Both the Dykstra–Parsons [19] and Stiles [35] methods are very cursory and, if used, they are normally followed by more extensive evaluations, usually by computer simulation.

For either the Dykstra–Parsons or Stiles methods, the permeabilities are arranged in descending order. For the Dykstra–Parsons method, the permeability variation is determined as described earlier. Two options are then possible: a program [36] for hand-held calculators can be used, or the graphical technique presented by Johnson [21] can be used. The fractional recovery, R, (see Figure 4.12 for example) expressed as a fraction of the oil in place when the waterflood is started, must be multiplied by the areal sweep efficiency (for example from Figure 4.9) to obtain the waterflood recovery.

For the Stiles technique, a program [37] for hand-held calculators is available or the procedure summarized in Table 4.7 can be used. A straightforward presentation of the Stiles method is in the text by Craft and Hawkins [51]. The fractional recovery obtained with the Stiles method is a fraction of the recoverable oil $(S_{op} - S_{or})$ that has been recovered at a given reservoir water cut. Since a water-oil ratio (WOR) is measured at surface conditions, the fractional water cut at reservoir conditions, f_w, is obtained (assuming $B_w = 1.0$) from:

$$f_w = \frac{WOR}{WOR + B_o} \qquad (4.25)$$

where B_o is the oil formation volume factor.

TABLE 4.7   Stiles Method of Calculating Waterflood Performance in Stratified Reservoirs

$$R = \frac{\left[k_j h_j + (c_t - c_j)\right]}{k_j h_t}$$

R = fraction of recoverable oil that has been produced
c_t = total capacity of formation (md-ft)
c_j = mid-ft which have been completely flooded with water
h_t = total net thickness of formation (ft)
h_j = total net thickness flooded (ft)
k_j = permeability of layer just flooded out

Reservoir conditions at P.W.

$$f_w = \frac{Mc_j}{[Mc_j + (c_t - c_j)]}$$

f_w = fractional flow of water

$$M = \frac{k_{rw}\mu_o}{k_{ro}\mu_w}$$

Surface conditions

$$f_w' = \frac{Ac_j}{[Ac_i + (c_t - c_j)]}$$

$$A = \frac{k_{rw}\mu_o B_o}{k_{ro}\mu_w B_w}$$

B_o = oil formation volume factor
B_w = water formation volume factor

From Reference 51.

4.12 PERFORMANCE EVALUATION

Monitoring waterflood performance is crucial to the success of the flood. From a reservoir engineering standpoint, the primary concerns are water injectivity and oil productivity. A few important factors related to these concerns will be summarized.

4.13 INJECTIVITY AND INJECTIVITY INDEX

Whereas productivity index was the ability of a well to produce hydro-carbons, injectivity index, I, in B/D/psi, is a measure of the ability of a well to accept fluids [51]:

$$I = \frac{q_{sc}}{P_{iwf} - P_e} \tag{4.26}$$

where q_{sc} is the flow rate in B/D at surface conditions, P_{iwf} is the flowing bottomhole pressure in psi, and P_e is the external pressure in psi. Some engineers express injectivity in terms of q_{sc}/p_{iwf} so that when injectivity is given, the reader is cautioned to understand what base pressure was intended. By dividing I by reservoir thickness, a specific injectivity index (specific to one well) can be obtained in B/D/psi/ft. In addition to expressing injectivity in terms of fluid injection rate in B/D, injectivity also is given as B/D/ac-ft and B/D/net ft of producing interval.

Values of injectivity depend on properties of the reservoir rock, well spacing, injection water quality, fluid-rock interactions, and pressure drop in the reservoir. Typical values of injectivity are in the range of 8–15 B/D/net ft or 0.75 – 1.0 B/D/net ac-ft. In waterflooding operations, water injection may begin into a reservoir produced by solution-gas-drive in which a mobile gas saturation exists, or injection may begin prior to the development of a mobile gas saturation. In the latter case, the system can be considered filled with liquid.

4.13.1 Injectivities for Various Flood Patterns

Analytical expressions for liquid-filled patterns were given by Muskat [50] and Deppe [38] for a mobility ratio of one (see Table 4.8). While these exact analytical solutions can be developed for steady-state pressure distributions, the equations in Table 4.8 cannot be used directly if the mobility is not one. However, the equations are useful in estimating injectivity in limiting conditions. For example, if k and μ are selected for oil at the connate water saturation, an estimate of initial injection rate can be obtained. Then if k and μ are selected for water at residual saturation, an estimate can be made of injectivity at 100% sweep. (These estimates can be useful when equipment is sized for a waterflood). If data on skin factor are available, suitable corrections [3, 52] can be inserted in the logarithm

TABLE 4.8 Injection Rates in Fully Developed Patterns at Unit Mobility Ratio

Direct Line Drive $\dfrac{d}{a} \geq 1$	Staggered Line Drive
$i = \dfrac{0.003541\,kh(\Delta p)}{\mu\left[\ln\left(\dfrac{a}{r_w}\right) + 1.571\dfrac{d}{a} - 1.838\right]}$	$i = \dfrac{0.003541\,kh(\Delta p)}{\mu\left[\ln\left(\dfrac{a}{r_w}\right) + 1.571\dfrac{d}{a} - 1.838\right]}$
Five-spot	Seven-spot
$i = \dfrac{0.003541\,kh(\Delta p)}{\mu\left[\ln\left(\dfrac{d}{r_w}\right) - 0.619\right]}$	$i = \dfrac{0.00472\,kh(\Delta p)}{\mu\left[\ln\left(\dfrac{d}{r_w}\right) - 0.569\right]}$
Nine-spot	Nine-spot
$i = \dfrac{0.003541\,kh(\Delta p)_{i,c}}{\left(\dfrac{1+R}{2+R}\right)\left[\ln\dfrac{d}{r_w} - 0.272\right]\mu}$	$i = \dfrac{0.00782\,kh(\Delta p)_{i,s}}{\left(\dfrac{3+R}{2+R}\left[\ln\dfrac{d}{r_w} - 0.272\right] - \dfrac{0.693}{2+R}\right)\mu}$

R = ratio of producing rate of corner well to side well
$(\Delta p)_{i,c}$ = pressure difference between injection well and corner well, and
$(\Delta p)_{i,s}$ = pressure difference between injection well and side well.

From References 50 and 38.
*Units in these equations are B/D, md, ft, psi, and cp.

term in the denominator in these equations. For unit mobility ratio, the injection rate will remain constant during the flood. If the mobility ratio is more than one, the injection rate increases as more water is injected; if the mobility ratio is less than one, the injection rate decreases. Figure 4.18 shows for different mobility ratios, the change in relative injectivity as the water bank extends radially from the injector [23]. Figure 4.19 shows, for different mobility ratios, the change in relative injectivity as a 40-acre 5-spot is swept [23].

For water injection into a depletion drive reservoir, several stages can describe the progress of the flood [53]. The first stage is the period of radial flow from the start of infection until interference of oil banks from adjacent injectors occurs. The second stage is the period from interference until fill-up of the pre-existing gas space; after fill-up, production response begins. The third stage is the period from fill-up to water breakthrough into the producing wells; water production begins at breakthrough. The fourth and final stage is the period from water breakthrough until floodout. For a 5-spot pattern, the injection rate during fill-up and to the time of interference can be estimated by [3]:

$$i = \frac{0.007082\,k_w \left(\dfrac{k_{ro}}{\mu_o}\right) h(p_{iwf} - p_{wf})}{\dfrac{1}{M}\ln\dfrac{r_f}{r_w} + \ln\dfrac{r_{ob}}{r_f}} \tag{4.27}$$

where oilfield units of B/D, md, cp, ft, and psi are used, and

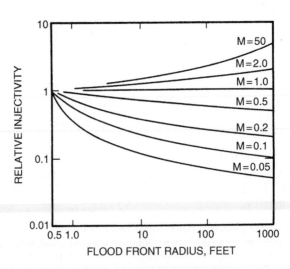

FIGURE 4.18 Change in injectivity at varying radial distances for different mobility ratios [23].

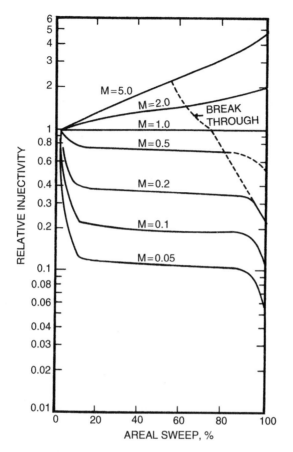

FIGURE 4.19   Change in injectivity as a function of area swept for different mobility ratios [23].

r_{ob} = radius of the oil bank, $r_w \leq r_{ob} \leq d/\sqrt{2}$
r_f = radius of the flood-front saturation

Both r_{ob} and r_f can be obtained by a material balance on the injected water:

$$W_i = \pi(r_f^2 - r_w^2)(\bar{S}_w - S_{iw})h\phi \qquad (4.28)$$

$$r_f = \sqrt{\frac{W_i}{\pi\phi h(\bar{S}_w - S_{iw})} + r_w^2} \qquad (4.29)$$

Since the volume of water injected to fill-up is equal to the volume of gas displaced by the oil bank as the initial gas saturation, S_{gi}, is reduced to the trapped-gas saturation, S_g, a material balance for r_{ob}, is:

$$r_{ob} = \sqrt{\frac{W_i}{\pi\phi h(S_{gi} - S_{gt})} + r_w^2} \qquad (4.30)$$

At interference, $r_{ob} = d/\sqrt{2}$, and Equation 4.30 can be used to compute the volume of water required to reach interference. Usually, fill-up occurs in a relatively short time after interference, and the volume of water injected at fill-up is:

$$W_{if} = 2d^2\phi h S_{gi} \qquad (4.31)$$

At fill-up, r_f is obtained by:

$$r_f = \sqrt{\frac{2d^2 S_{gi}}{\pi(\bar{S}_w - S_{iw})} + r_w^2} \qquad (4.32)$$

After interference, the equation in Table 4.8 for the 5-spot pattern can be used to estimate injectivity. Additional details of estimating injectivity can be found in several good texts on this subject [3, 52].

4.13.2 Monitoring Injectivity

Injection well performance can be analyzed and monitored by several means. During and after a period of injection, the pressure transient methods discussed earlier can be used. Additionally, several bookkeeping methods of monitoring injection rates and pressures are quite useful.

Hearn [39] recently proposed a method to analyze injection well pressure and rate data. Permeability is obtained from the slope of a plot of $\Delta p/q_w$ versus the logarithm of cumulative water injected. However, the method can only be used during the initial injection period. After fill-up, $\Delta p/q$, the reciprocal of injectivity index, will cease to be a function of cumulative water injected unless the well experiences damage or is stimulated. In these cases, the plots suggested by Hall [40] are convenient for analysis of the data.

A Hall plot is a graph of cumulative pressure-time versus cumulative water injection. Such plots are useful in observing injection well plugging or any beneficial results of stimulation procedures. An improvement in injectivity is indicated if the slope decreases, whereas plugging is suspected if the slope steepens. Figure 4.20 shows an improvement in water injectivity that resulted from a surfactant treatment [41].

The reciprocal of the Hall plot slope is the injectivity index in bbl/D/psi. Effective pressures are obtained by subtracting the static reservoir pressure from the flowing bottomhole pressures [40]:

$$\text{effective pressure} = P_{iwf} - \bar{p} = (p_{wh} + 0.45D - \Delta p_t) - \bar{p} \qquad (4.33)$$

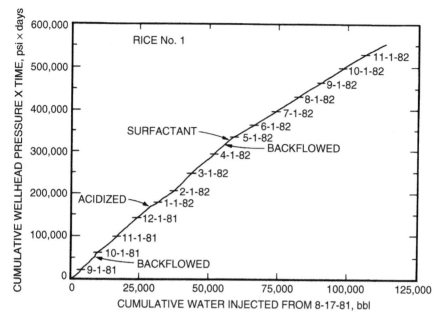

FIGURE 4.20   Hall plot [40].

where p_{wh} is the wellhead pressure, 0.45 is the hydrostatic pressure gradient in psi/ft, D is the depth to the mid-point of the reservoir, and Δp_t is the pressure drop in the tubing. Although the Hall method assumes that only the wellhead pressure changes with cumulative water injected, the effective pressure drop should be used if permeability capacity, or transmissibility changes, are desired [40]. If the difference between reservoir pressure and the hydrostatic head is more than 15% of the wellhead pressure, serious quantitative errors can result when wellhead pressures are used in the construction of Hall plots; if the difference is less than 15%, only slight errors are caused [54].

Earlougher has presented a modified version of the Hall technique in which the slope, m', is defined as [54]:

$$m' = \frac{141.2\mu}{kh}\left[\ln\left(\frac{r_e}{r_w}\right) + s\right] \tag{4.34}$$

The improvement in overall flow efficiency, E_f, can be obtained from [54]:

$$\frac{E_f \text{ after}}{E_f \text{ before}} = \frac{m' \text{ before}}{m' \text{ after}} = \frac{I \text{ after}}{I \text{ before}} \tag{4.35}$$

where the injectivity index, I, or Hall slopes are calculated using the pressure difference between the flowing bottomhole pressure and the formation pressure.

4.13.3 Production Curves

Plots of waterflood injection and production performance can be presented in a number of ways. For the history of the project, water injection rate, oil production rate, and water-oil ratio or water cut can be plotted vs. time (usually months). The actual water injection and oil production rates can be compared to the predicted rates on a time basis.

Future oil production and ultimate recovery are often extrapolated from graphical methods. One of the more popular methods is a plot of the WOR on a log scale vs. cumulative oil production on a linear scale or a linear plot of the fractional water cut (or percent water produced) vs. cumulative oil produced. Alternatively, the oil-water ratio can be plotted on a log scale vs. the cumulative production on a linear scale. One of the purposes of these plots is to predict the ultimate oil recovery by extrapolating the curve to some economic limit at which time it becomes no longer profitable to continue the flood. If the operating methods remain relatively unchanged, a method [42] has been proposed for a fully developed waterflood that permits an easy extrapolation of recovery to a given water cut. This latter method consists of a linear plot of E_R, fractional recovery of oil in place, vs. the term $-\{[(1/f_w) - 1] - (1/f_w)\}$. This method also provides an estimate of water-oil relative permeabilities.

TABLE 4.9   Waterflood Parameters

Waterflood Residual Oil Saturations Measured in Core Test Samples		
	Sandstone	Carbonate Rock
Mean average S_{or}, % pore space	27.7	26.2
Median average S_{or}, % pore space	26.6	25.2
Number of core samples tested	316	108
Number of source reservoirs	75	20
Standard deviation, % pore space	8.76	8.84

Effect of Lithology and K_a on the end Points of oil/water relative permeability curves			
Permeability Group range of K_a, md	Low (1 to 10)	Medium (11 to 100)	High (101 to 2,000)
Sandstone			
Median k_{rw}^* at S_{or}	0.065	0.133	0.256
Median k_{rw}^{**} at S_{or}	0.033	0.095	0.210
Number of core samples tested	30	213	143
Carbonate rock			
Median k_{rw}^* at S_{or}	0.211	0.357	0.492
Median k_{rw}^{**} at S_{or}	0.179	0.303	0.428
Number of core samples tested	33	45	24

From reference 43.
*Expressed as fraction of k_0 at connate water saturation.
**Expressed as fraction of K_a.

4.13.4 Waterflood Parameters

Important parameters in waterflood operations are the water residual oil saturation, S_{or}, and the relative permeability to water, k_{rw}. A statistical study of these parameters, as well as peak oil rates, was provided by Felsenthal [43]. Data on S_{or} and k_{rw} from core data are listed in Table 4.9. Endpoint k_{rw} values were higher in carbonates than in sandstones; for a given lithology, k_{rw} decreased as the absolute permeability decreased.

References

[1] Langnes, G. L., Robertson, J. O., Jr., and Chilingar, G.V., *Secondary Recovery and Carbonate Reservoirs*, American Elsevier Publishing Co., New York (1972).

[2] *Improved Oil Recovery*, Interstate Oil Compact Commission, Oklahoma City (March 1983).

[3] Willhite, G. P., *Waterflooding*, SPE, Richardson, TX (1986).

[4] Singh, S. P., and Kiel, O. G., "Waterflood Design (Pattern, Rate, and Timing)," paper SPE 10024 presented at the SPE 1982 Intl. Pet. Exhibition & Tech. Symposium, Bejing, China, March 18–26.

[5] Muskat, M., and Wyckoff, R. D., "A Theoretical Analysis of Waterflooding Networks," *Trans.*, AIME, Vol. 107 (1934), pp. 62–76.

[6] Dyes, A. B., Caudle, B. H., and Erickson, R. A., "Oil Production After Breakthrough-As Influenced by Mobility Ratio," *Trans.*, AIME, Vol. 201 (1954), pp. 81–86.

[7] Craig, F. F., Jr., Geffen, T. M., and Morse, R. A., "Oil Recovery Performance of Pattern Gas or Water Injection Operations from Model Tests," *J. Pet. Tech.* (Jan. 1955), pp. 7–15; *Trans.*, AIME, Vol. 204.

[8] Prats, M., et al., "Prediction of Injection Rate and Production History for Multifluid Five-Spot Floods," *J. Pet. Tech.* (May 1959), pp. 98–105; *Trans.*, AIME, Vol. 216.

[9] Caudle, B. H., and Witte, M. D., "Production Potential Changes During Sweep-Out in a Five-Spot Pattern," *Trans.*, AIME, Vol. 216 (1959), pp. 446–448.

[10] Habermann, B., "The Efficiency of Miscible Displacement As a Function of Mobility Ratio," *Trans.*, AIME, Vol. 219 (1960), pp. 264–272.

[11] Kimbler, O. K., Caudle, B. H., and Cooper, H. E., Jr., "Areal Sweep-out Behavior in a Nine-Spot Injection Pattern," *J. Pet. Tech.* (Feb. 1964), pp. 199–202.

[12] Caudle, B. H., Erickson, R. A., and Slobod, R. L., "The Encroachment of Injected Fluids Beyond the Normal Well Patterns," *Trans.*, AIME, Vol. 204 (1955), pp. 79–85.

[13] Prats, M., Hazebroek, P., and Allen, E. E., "Effect of Off-Pattern Wells on the Performance of a Five-Spot Waterflood," *J. Pet. Tech.* (Feb. 1962), pp. 173–178.

[14] Prats, M., "The Breakthrough Sweep Efficiency of the Staggered Line Drive," *Trans.*, AIME, Vol. 207 (1956), pp. 361–362.

[15] Crawford, P. B., "Factors Affecting Waterflood Pattern Performance and Selection," *J. Pet. Tech.* (Dec. 1960), pp. 11–15.

[16] Fassihi, M. R., "New Correlations for Calculation of Vertical Coverage and Areal Sweep Efficiency," *SPE Reservoir Engineering* (Nov. 1986), pp. 604–606.

[17] Claridge, E. L., "Prediction of Recovery in Unstable Miscible Flooding," *Soc. Pet. Eng. J.* (April 1972), pp. 143–155.

[18] Schmalz, J. P., and Rahme, H. D., "The Variation of waterflood Performance With Variation in Permeability Profile," *Prod. Monthly*, Vol. 15, No. 9 (Sept. 1950), pp. 9–12.

[19] Dykstra, H., and Parsons, R. L., "The Prediction of Oil Recovery by Water-flood," *Secondary Recovery of Oil in the United States*, API, Dallas (1950) second edition, pp. 160–174.

[20] Warren, J. E., and Cosgrove, J. J., "Prediction of Waterflood Behavior in A Stratified System," *Soc. Pet. Eng. J.* (June 1964), pp. 149–157.

[21] Johnson, C. E., Jr., "Prediction of Oil Recovery by Water Flood-A Simplified Graphical Treatment of the Dykstra–Parsons Method," *Trans.*, AIME, Vol. 207 (1956), pp. 345–346.

[22] deSouza, A. O., and Bridham, W. E., "A Study on Dykstra–Parsons Curves," Tech. Report 29, Stanford University, Palo Alto (1981).

[23] Craig, F. F., Jr., "Effect or Reservoir Description on Performance Predictions," *J. Pet. Tech.* (Oct. 1970), pp. 1239–1245.

[24] Reznik, A. A., Emick, R. M., and Panvelker, S. B., "An Analytical Extension of the Dykstra–Parsons Vertical Stratification Discrete Solution to a Continuous, Real-Time Basis," *Soc. Pet. Eng. J.* (Dec. 1984), pp. 643–656.

[25] Root, P. J., and Skiba, F. F., "Cross flow Effects Duringan Idealized Process in A Stratified Reservoir," *Soc. Pet. Eng. J.* (Sept. 1965), pp. 229–238.

[26] Goddin, C. S., Jr., et al., "A Numerical Study of Waterflood Performance in a Stratified System With Crossflow," *J. Pet. Tech.* (June 1966), pp. 765–771.

[27] El-Khatib, N., "The Effect of Crossflow on Waterflooding of Stratified Reservoirs," *Soc. Pet. Eng. J.* (April 1985), pp. 291–302.

[28] Collins, H. N., and Wang, S. T., "Discussion of the Effect of Crossflow on Waterflooding of Stratified Reservoirs," *Soc. Pet. Eng. J.* (Aug. 1985), p. 614.

[29] El-Khatib, N., "Author's Reply to Discussion of the Effect of Cross flow on Waterflood of Stratified Reservoirs," *Soc. Pet. Eng. J* (Aug. 1985), p. 614.

[30] Collins, H. N., and Wang, S. T., "Further Discussion of the Effect of Crossflow on Water-flooding of Stratified Reservoirs," *SPE Reservoir Eng.* (Jan. 1986) p. 73.

[31] El-Khatib, N., "Author's Reply to Further Discussion of the Effect of Crossflow on Water-flooding of Stratified Reservoirs," *SPE Reservoir Eng.* (Jan 1986), pp. 74–75.

[32] Hirasaki, G. J., Morra, F., and Willhite, G. P., "Estimation of Reservoir Hetrogeneity from Waterflood Performance," SPE 13415, unsolicited paper.

[33] Hirasaki, G. J., "Properties of Log-Normal Permeability Distribution for Stratified Reservoirs," SPE 13416, unsolicited paper.

[34] Higgins, R. V., ad Leighton, A. J., "A Computer Method to Calculate Two-Phase Flow in Any Irregularly Bounded Porous Medium," *J. Pet Tech.* (June 1962), pp. 679–683; *Trans.* AIME, Vol. 225.

[35] Stiles, W. E., "Use of Permeability Distribution in Water Flood Calculations," *Trans.*, AIME, Vol. 186 (1949), pp. 9–13.

[36] Garb, F. A., "Waterflood Calculations for Hand-held Computers. Part 8—Using Dykstra-Parsons Methods to Evaluate Stratified Reservoirs," *World Oil* (July 1980), pp. 155–160.

[37] Garb, F. A., "Waterflood Calculations for Hand-held Computers. Part 7—Evaluating Flood Performance Using the Stiles Technique," *World Oil* (June 1980), pp. 205–210.

[38] Deppe, J. C., "Injection Rates-The Effect of Mobility Ratio, Area Swept, and Pattern," *Soc. pet. Eng. J.* (June 1961), pp. 81–91; *Trans.*, AIME, Vol. 222.

[39] Hearn, C. L., "Method Analyzes Injection Well Pressure and Rate Data," *Oil & Gas J.* (April 18, 1983), pp. 117–120.

[40] Hall, H. N., "How to Analyze Waterflood Injection Well Performance," *World Oil* (Oct. 1963), pp. 128–130.

[41] Martin, F. D., "Injectivity Improvement in the Grayburg Formation at a Waterflood in Lea Country, NM," paper SPE 12599 presented at the 1984 Permian Basin Oil & Gas Recovery Conference, Midland, March 8–9.

[42] Ershaghi, I., and Omoregie, O., "A Method for Extrapolation of Cut vs. Recovery Curves," *J. Pet. Tech.* (Feb. 1978), pp. 203–204.

[43] Felsenthal, M., "A Statistical Study of Some Waterflood Parameters," *J. Pet. Tech.* (Oct. 1979), pp. 1303–1304.

[44] Craig, F. F., Jr., "The Reservoir Engineering Aspects of Waterflooding," Monograph Series, SPE, Dallas, Vol. 3 (1971).

[45] Leverett, M. C., "Capillary Behavior in Porous Solids," *Trans.*, AIME, Vol. 142 (1941), pp. 159–172.

[46] Buckley, S. E., and Leverett, M. C., "Mechanism of Fluid Displacement in Sands," *Trans.*, AIME, Vol. 146 (1942), pp. 107–116.

[47] Welge, H. J., "A Simplified Method for Computing Oil Recovery by Gas or Water Drive," *Trans.*, AIME, Vol. 195 (1952), pp. 91–98.

[48] "Statistical Analysis of Crude Oil Recovery and Recovery Efficiency," API BUL D14, second edition, API Prod. Dept., Dallas (Apr. 1984).

[49] Doscher, T. M., "Statistical Analysis Shows Crude-Oil Recovery," *Oil & Gas J.* (Oct. 29, 1984), pp. 61–63.

[50] Muskat, M., *Physical Principles of Oil Production*, McGraw-Hill Book Co., Inc., New York (1949).

[51] Craft, B. C., and Hawkins, M. F., *Applied Petroleum Reservoir Engineering*, Prentice-Hall, Inc., Englewood Cliffs (1959).

[52] Slider, H. C., *Worldwide Practical Petroleum Reservoir Engineering Methods*, PenWell Pub. Co., Tulsa (1983).

[53] Timmerman, E. H., *Practical Reservoir Engineering*, PennWell Books, Tulsa (1982).

[54] Earlougher, R. C., Jr., *Advances in Well Test Analysis*, Monograph Series, SPE, Dallas (1977), Vol. 5.

Enhanced Oil Recovery Methods

5.1 DEFINITION

A general schematic of the enhanced oil recovery (EOR) process is depicted in Figure 5.1. The more common techniques that are currently being investigated include:

Enhanced Oil Recovery
Chemical Oil Recovery or Chemical Flooding
  Polymer-augmented waterflooding
  Alkaline or caustic flooding
  Surfactant flooding
  —Low tension waterflooding
  —Micellar/polymer (microemulsion) flooding

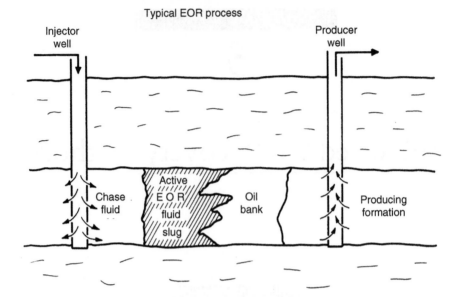

FIGURE 5.1   General schematic of enhanced oil recovery.

Hydrocarbon or Gas Injection
   Miscible solvent (LPG or propane)
   Enriched gas drive
   High-pressure gas drive
   Carbon dioxide flooding
   Flue gas
   Inert gas (nitrogen)
Thermal Recovery
   Steamflooding
   In-situ combustion

These procedures are discussed in several texts on the subject [1–5, 32]. Two studies by the National Petroleum Council [6, 7] and several papers summarizing the later study are available [8–11]. The extensive literature on enhanced recovery will not be cited, and the reader is referred to Reference 12 which provides numerous citations and is the basis of the following discussion.

5.2 CHEMICAL FLOODING

Chemical oil recovery methods include polymer, surfactant/polymer (variations are called micellar-polymer, microemulsion, or low tension waterflooding), and alkaline (or caustic) flooding. All of these methods involve mixing chemicals (and sometimes other substances) in water prior to injection. Therefore, these methods require conditions that are very favorable for water injection: low-to-moderate oil viscosities, and moderate-to-high permeabilities. Hence, chemical flooding is used for oils that are more viscous than those oils recovered by gas injection methods but less viscous than oils that can be economically recovered by thermal methods. Reservoir permeabilities for chemical flood conditions need to be higher than for the gas injection methods, but not as high as for thermal methods. Since lower mobility fluids are usually injected in chemical floods, adequate injectivity is required. If previously waterflooded, the chemical flood candidate should have responded favorably by developing an oil bank. Generally, active water-drive reservoirs should be avoided because of the potential for low remaining oil saturations. Reservoirs with gas caps are ordinarily avoided since mobilized oil might resaturate the gas cap. Formations with high clay contents are undesirable since the clays increase adsorption of the injected chemicals. In most cases, reservoir brines of moderate salinity with low amounts of divalent ions are preferred since high concentrations interact unfavorably with the chemicals that are injected.

5.2.1 Polymer-Augmented Waterflooding

High mobility ratios cause poor displacement and sweep efficiencies, and result in early breakthrough of injected water. By reducing the mobility of water, water breakthrough can be delayed by improving the displacement, areal, and vertical sweep efficiencies; therefore more oil can be recovered at any given water cut. Thus, the ultimate oil recovery at a given economic limit may be 4%–10% higher with a mobility-controlled flood than with plain water. Additionally, the displacement is more efficient in that less injection water is required to produce a given amount of oil.

The need to control or reduce the mobility of water led to the advent of polymer flooding or polymer-augmented waterflooding. Polymer flooding is viewed as an improved waterflooding technique since it does not ordinarily recover residual oil that has been trapped in pore spaces and isolated by water. However, polymer flooding can produce additional oil over that obtained from waterflooding by improving the displacement efficiency and increasing the volume of reservoir that is contacted. Dilute aqueous solutions of water-soluble polymers have the ability to reduce the mobility of water in a reservoir thereby improving the efficiency of the flood. Partially hydrolyzed polyacrylamides (HPAM) and xanthan gum (XG) polymers both reduce the mobility of water by increasing viscosity. In addition, HPAM can alter the flow path by reducing the permeability of the formation to water. The reduction in permeability to water that is achieved with HPAM solution can be fairly permanent while the permeability to oil can remain relatively unchanged. The resistance factor is a term that is commonly used to indicate the resistance to flow that is encountered by a polymer solution as compared to the flow of plain water. For example, if a resistance factor of 10 is observed, it is 10 times more difficult for the polymer solution to flow through the system, or the mobility of water is reduced 10-fold. Since water has a viscosity of about 1 cp, the polymer solution, in this case, would flow through the porous system as though it had an apparent or effective viscosity of 10 cp even though a viscosity measured in a viscometer could be considerably lower.

The improvement in areal sweep efficiency resulting from polymer treatment can be estimated from Figure 4.9. For example, if the mobility ratio for a waterflood with a 5-spot pattern is 5, the areal sweep efficiency is 52% at breakthrough. If the economic limit is a producing water-oil ratio of 100:1 ($f_w \cong 100/101 = 0.99$), the sweep efficiency at floodout is about 97%. If the polymer solution results in the mobility ratio being lowered to 2, sweep efficiencies are 60% at breakthrough and 100% at the same economic water-oil ratio.

A simplified approach to qualitatively observing the improvement with polymers in a stratified system is illustrated in Figure 4.12. For example, if the permeability variation is 0.7, the waterflood mobility ratio is 5, and the initial water saturation is 0.3, the fractional recovery of oil-in-place can be estimated. From the plot, $R(1 - 0.4S_w) = 0.29$, and the fractional recovery, R, is $0.29/[1 - (0.4)(0.3)] = 0.33$. This R needs to be multiplied by the areal sweep efficiency of 0.97 to yield a recovery of 32% of the oil-in-place. If polymers again reduce the mobility ratio to 2 (and if no improvement in permeability variation occurs), a fractional recovery of 0.375 is obtained. Since the areal sweep with the polymer flood is 100%, a recovery of 37.5% of the oil-in-place is estimated. Thus the improvement with polymers is estimated at 0.375–0.32 or 5.5% of the oil-in-place. If the flow distribution with polymer solution lowered the permeability variation (which is not likely), the incremental production could be higher. These calculations are gross oversimplifications of actual conditions and only serve as a tool to show that reducing mobility ratio with polymers can improve the sweep efficiencies.

A properly sized polymer treatment may require the injection of 15%–25% of a reservoir pore volume; polymer concentrations may normally range from 250 to 2,000 mg/L. For very large field projects, millions of pounds of polymer may be injected over a 1–2 year period of time; the project then reverts to a normal waterflood. The polymer flooding literature was reviewed in the late 1970s [13]. Recommendations on the design of polymer floods were recently made available [14].

5.2.2 Variations in the Use of Polymers

In-Situ Polymerization A system is available in which acrylamide monomer is injected and polymerized in the reservoir. Both injection wells and producing wells have been treated.

Crosslinked or Gelled Polymers Several methods are available for diverting the flow of water in reservoirs with high permeability zones or fracture systems. Some methods are only effective near the injection well while others claim the treatment can be effective at some depth into the reservoir. Both producing wells and injection wells can be treated.

One method is the aluminum citrate process which consists of the injection of a slug of HPAM polymer solution, aluminum ion chelated with citrate ion, and a second slug of polymer. Some of the polymer in the first slug adsorbs or is retained on the surfaces of the reservoir. The aluminum ion attaches to the adsorbed polymer and acts as a bridge to the second polymer layer. This sequence is repeated until the desired layering is achieved. The transport of aluminum ions through the reservoir may be a problem in certain cases, so the effects of the treatment may be limited to near the wellbore.

Another method is based on the reduction of chromium ions to permit the crosslinking of HPAM or XG polymer molecules. A polymer slug containing Cr^{+6} is injected followed by a slug of polymer containing a reducing agent. When the Cr^{+6} is reduced to Cr^{+3}, a gel is formed with the polymer. The amount of permeability reduction is controlled by the number of times each slug is injected, the size of each slug, or the concentrations used. An alternate treatment involves placing a plain water pad between the first and second polymer slugs.

In another variation of the above two methods for HPAM, a cationic polymer is injected first. Since reservoir surfaces are often negatively charged, the cationic polymer is highly adsorbed. When the foregoing sequential treatments are injected, there is a strong attraction between the adsorbed cationic polymer and the anionic polymers that follow.

Polymer concentrations used in these variations are normally low, on the order of 250 mg/L. With low molecular weight polymers or if a very stiff gel is desired, polymer concentrations of 1–1.5% are common. The type of polymers are similar to those used in conventional polymer flooding, but the products used for gelation command a higher price.

Methods developed recently, especially for fracture treatments, include Cr^{3+} (acetate)-polyacrylamide, collordal silica, and resorcinol-formaldehyde.

5.2.3 Surfactant and Alkaline Flooding

Both alkaline flooding and surfactant flooding improve oil recovery by lowering the interfacial tension between crude oil and the displacing water. With alkaline flooding, the surfactants are generated in situ when the alkaline materials react with crude oil–this technique is normally only viable when the crude oil contains sufficient amount of organic acids to produce natural surfactants. Other possible mechanisms with the caustic materials include emulsification of the oil and alteration in the preferential wettability of the reservoir rock.

With surfactant flooding, surface-active agents are mixed with other compounds (such as alcohol and salt) in water and injected to mobilize the crude oil. Polymer-thickened water is then injected to push the mobilized oil-water bank to the producing wells. Water-soluble polymers can be used in a similar fashion with alkaline flooding. For micellar/polymer flooding, the concentration of polymer used may be similar to the value given for polymer flooding, but the volume of polymer solution may be increased to 50% or more of a reservoir pore volume.

Alkaline Flooding   Alkaline or caustic flooding consists of injecting aqueous solutions of sodium hydroxide, sodium carbonate, sodium silicate or potassium hydroxide. The alkaline chemicals react with organic acids in certain crude oils to produce surfactants in situ that dramatically lower the interfacial tension between water and oil. The alkaline agents also react with the reservoir rock surfaces to alter wettability-either from oil-wet to water-wet, or vice versa. Other mechanisms include emulsification and entrainment of oil or emulsification and entrapment of oil to aid in mobility control. Since an early patent in the 1920s described the use of caustic for improved recovery of oil, much research and some field tests have been conducted. Slug size of the alkaline solution is often 10%–15% PV, concentrations of the alkaline chemical are normally 0.2% to 5%. Recent tests are using large amounts of relatively high concentrations. A preflush of fresh or softened water often precedes the alkaline slug, and a drive fluid (either water or polymer-thickened water) follows the alkaline slug.

Surfactant/Polymer Flooding   Surfactant use for oil recovery is not a recent development. Patents in the late 1920s and early 1930s proposed the use of low concentrations of detergents to reduce the interfacial tension between water and oil. To overcome the slow rate of advance of the detergent, Taber [15] proposed very high concentrations ($\sim 10\%$) of detergent in aqueous solution.

During the late 1950s and early 1960s, several different present-day methods of using surfactants for enhanced recovery were developed. A review of these methods is beyond the scope of this chapter and is available in the literature [16–19]. In some systems, a small slug (> about 5% PV) was proposed that included a high concentration of surfactant (normally 5%–10%). In many cases, the microemulsion includes surfactant, hydrocarbon, water, an electrolyte (salt), and a cosolvent (usually an alcohol). These methods ordinarily used a slug (30%–50% PV) of polymer-thickened water to provide mobility control in displacing the surfactant and oil-water bank to the producing wells. The polymers used are the same as those discussed in the previous section. In most cases, low-cost petroleum sulfonates or blends with other surfactants have been used. Intermediate surfactant concentrations and low concentration systems (low tension waterflooding) have also been proposed. The lower surfactant concentration systems may or may not contain polymer in the surfactant slug, but will utilize a larger slug (30%–100% PV) of polymer solution.

Alkaline/Surfactant/Polymer Flooding A recent development uses a combination of chemicals to lower process costs by lowering injection cost and reducing surfactant adsorption. These mixtures, termed alkaline/surfactant/polymer (ASP), permit the injection of larger slugs of injectant because of the lower cost.

5.3 GAS INJECTION METHODS

5.3.1 Hydrocarbon Miscible Flooding

Gas injection is certainly one of the oldest methods utilized by engineers to improve recovery, and its use has increased recently, although most of the new expansion has been coming from the nonhydrocarbon gases [20]. Because of the increasing interest in CO_2 and nitrogen or flue gas methods, they are separated from the hydrocarbon miscible techniques.

Hydrocarbon miscible flooding can be subdivided further into three distinct methods, and field trials or extensive operations have been conducted in all of them. For LPG slug or solvent flooding, enriched (condensing) gas drive and high pressure (vaporizing) gas drive, a range of pressures (and therefore, depths) are needed to achieve miscibility in the systems.

Unless the reservoir characteristics were favorable, early breakthrough and bypassing of large quantities of oil have plagued many of the field projects. In addition, the hydrocarbons needed for the processes are valuable, and there is increasing reluctance to inject them back into the ground when there is some question about the percentage that will be recovered the second time around. Therefore, in the U.S. in recent years the emphasis has been shifting to less valuable nonhydrocarbon gases such as CO_2, nitrogen, and flue gases. Although nitrogen and flue gases do not recover oil as well as the hydrocarbon gases (or liquids), the overall economics may be somewhat more favorable.

5.3.2 Nitrogen and Flue Gas Flooding

As previously mentioned, nitrogen and flue gas (about 87% N2 and 12% CO_2) are sometimes used in place of hydrocarbon gases because of economics. Nitrogen also competes with CO_2 in some situations for the same reason. The economic appeal of nitrogen stems not only from its lower cost on a standard Mcf basis, but also because its compressibility is much lower.

Thus, for a given quantity at standard conditions, nitrogen will occupy much more space at reservoir pressures than CO_2 or even methane at the same conditions. However, both nitrogen or flue gas are inferior to hydrocarbon gases (and much inferior to CO_2) from an oil recovery point of view. Nitrogen has a lower viscosity and poor solubility in oil and requires a much higher pressure to generate or develop miscibility. The increase in the required pressure is significant compared to methane and very large (4–5 times) when compared to CO_2. Therefore, nitrogen will not reduce the displacement efficiency too much when used as a chase gas for methane, but it can cause a significant drop in the effectiveness of a CO_2 flood if the reservoir pressures are geared to the miscibility requirements for CO_2 displacements. Indeed, even methane counts as a desirable "light end" or "intermediate" in nitrogen flooding, but methane is quite deleterious to the achievement of miscibility in CO_2 flooding at modest pressures.

5.3.3 Carbon Dioxide Flooding

CO_2 is effective for recovery of oil for a number of reasons. In general, carbon dioxide is very soluble in crude oils at reservoir pressures; therefore, it swells the net volume of oil and reduces its viscosity even before miscibility is achieved by the vaporizing gas drive mechanism. As miscibility is approached, both the oil phase and the CO_2 phase (which contains many of the oil's intermediate components) can flow together because of the low interfacial tension and the relative increase in the total volumes of the combined CO_2 and oil phases compared to the water phase. However, the generation of miscibility between the oil and CO_2 is still the most important mechanism, and it will occur in CO_2- crude oil systems as long as the pressure is high enough. This so-called "minimum miscibility pressure" or MMP has been the target of several laboratory investigations and is no longer a mystery. The 1976 NPC report [6] showed that there is a rough correlation between the API gravity and the required MMP, and that the MMP increased with temperature. Some workers have shown that a better correlation is obtained with the molecular weight of the C_5^+ fraction of the oil than with the API gravity. In general the recent work shows that the required pressure must be high enough to achieve a minimum density in the CO_2 phase [21, 22]. At this minimum density, which varies with the oil composition, the CO_2 becomes a good solvent for the oil, especially the intermediate hydrocarbons, and the required miscibility can be generated or developed to provide the efficient displacement normally observed with CO_2. Therefore, at higher temperatures, the higher pressures are needed only to increase the CO_2 density to the same value as observed for the MMP at the lower temperature. Figure 5.2 shows the variation of minimum miscibility pressure with temperature and oil composition [23].

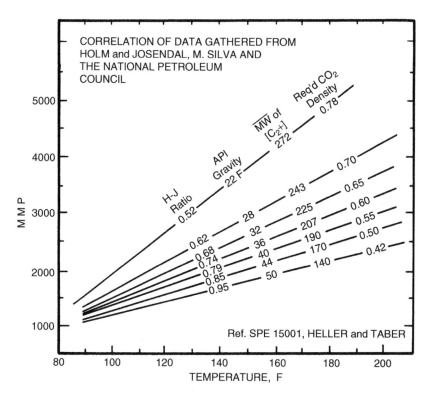

FIGURE 5.2 Correlations for CO_2 minimum miscibility pressure [23].

Although the mechanism for CO_2 flooding appears to be the same as that for hydrocarbon miscible floods, CO_2 floods may give better recoveries even if both systems are above their required miscibility pressures, especially in tertiary floods. Compared to hydrocarbons, CO_2 has a much higher solubility in water, and it has been observed in laboratory experiments to diffuse through the water phase to swell bypassed oil until the oil is mobile. Thus, not only are the oil and depth screening criteria easier to meet in CO_2 flooding, but the ultimate recovery may be better than with hydrocarbons when above the MMP. It must be noted, however that this conjecture has not been proved by rigorous and directly comparable experiments.

5.4 THERMAL RECOVERY

5.4.1 In-Situ Combustion

The theory and practice of in-situ combustion or fireflooding is covered comprehensively in the recent SPE monograph on thermal recovery by

Prats [4]. In addition, the continuing evolution of screening criteria for fire-flooding [24, 25] and steamflooding [26] have been reviewed and evaluated by Chu. A recent appraisal of in-situ combustion was provided by White [27] and the status of oxygen fireflooding was provided by Garon [28].

Part of the appeal of fireflooding comes from the fact that it uses the world's cheapest and most plentiful fluids for injection: air and water. However, significant amounts of fuel must be burned, both above the ground to compress the air, and below ground in the combustion process. Fortunately, the worst part of the crude oil is burned; the lighter ends are carried forward in advance of the burning zone to upgrade the crude oil.

5.4.2 Steam Flooding

Of all of the enhanced oil recovery processes currently available, only the steam drive (steamflooding) process is routinely used on a commercial basis. In the United States, a majority of the field testing with this process has occurred in California, where many of the shallow, high-oil-saturation reservoirs are good candidates for thermal recovery. These reservoirs contain high-viscosity crude oils that are difficult to mobilize by methods other than thermal recovery.

In the steam drive process, steam is continuously introduced into injection wells to reduce the viscosity of heavy oil and provide a driving force to move the more mobile oil towards the producing wells. In typical steam drive projects, the injected fluid at the surface may contain about 80% steam and 20% water (80% quality) [6]. When steam is injected into the reservoir, heat is transferred to the oil-bearing formation, the reservoir fluids, and some of the adjacent cap and base rock. As a result, some of the steam condenses to yield a mixture of steam and hot water flowing through the reservoir.

The steam drive may work by driving the water and oil to form an oil bank ahead of the steamed zone. Ideally this oil bank remains in front, increasing in size until it is produced by the wells offsetting the injector. However, in many cases, the steam flows over the oil and transfers heat to the oil by conduction. Oil at the interface is lowered in viscosity and dragged along with the steam to the producing wells. Recoverability is increased because the steam (heat) lowers the oil viscosity and improves oil mobility. As the more mobile oil is displaced the steam zone expands vertically, and the steam-oil interface is maintained. This process is energy-intensive since it requires the use of a significant fraction (25%–40%) of the energy in the produced petroleum for the generation of steam.

In steamflooding, the rate of steam injection is initially high to minimize heat losses to the cap and base rock. Because of reservoir heterogeneities and gravity segregation of the condensed water from the steam vapor, a highly permeable and relatively oil-free channel often develops between injector and producer. Many times this channel occurs near the top of the oil-bearing rock, and much of the injected heat is conducted to the caprock as heat loss rather than being conducted to oil-bearing sand where the heat is needed. In addition, the steam cannot displace oil efficiently since little oil is left in the channel. Consequently, neither the gas drive from the steam vapor nor the convective heat transfer mechanisms work as efficiently as desired. As a result, injected steam will tend to break through prematurely into the offset producing wells without sweeping the entire heated interval.

5.5 TECHNICAL SCREENING GUIDES

In some instances, only one type of enhanced recovery technique is applicable for a specific field condition but, in many instances, more than one technique is possible. The selection of the most appropriate process is facilitated by matching reservoir and fluid properties to the requirements necessary for the individual EOR techniques. A summary of the technical screening guides for the more common EOR processes is given in Table 5.1. A distinction is made between the oil properties and reservoir characteristics that are required for each process. Generally, steamflooding is applicable for very viscous oils in relatively shallow formations. On the other extreme, CO_2 and hydrocarbon miscible flooding work best with very light oils at depths that are great enough for miscibility to be achieved. Both steamflooding and in-situ combustion require fairly high permeability reservoirs. Chemical flooding processes (polymer, alkaline, or surfactant) are applicable in low to medium viscosity oils; depth is not a major consideration except, at great depths, the higher temperature may present problems in the degradation or consumption of some of the chemicals.

Screening guides or criteria are among the first items considered when a petroleum engineer evaluates a candidate reservoir for enhanced oil recovery. A source often quoted for screening criteria is the 1976 National Petroleum Council (NPC) report on Enhanced Recovery [6], which was revised by the NPC in 1984 [7]. Both reports list criteria for six enhanced recovery methods.

Some reservoir considerations apply to all enhanced recovery methods. Because drilling costs increase markedly with depth, shallow reservoirs are preferred, as long as all necessary criteria are met. For the most part, reservoirs that have extensive fractures, gross heterogeneities, thief zones, or are highly faulted should be avoided. Ideally, relatively uniform reservoirs with reasonable oil saturations, minimum shale stringers, and good areal extent are desired.

Implementation of enhanced recovery projects is expensive, time-consuming, and people-intensive. Substantial costs are often involved in the assessment of reservoir quality, the amount of oil that is potentially recoverable, laboratory work associated with the EOR process, computer

TABLE 5.1 Summary of Screening Criteria for Enhanced Recovery Methods

	Oil Properties			Reservoir Characteristics					
	Gravity °API	Viscosity cp	Composition	Oil Saturation	Formation Type	Net Thickness ft	Average Permeability md	Depth ft	Temperature °F
Gas injection methods									
Hydrocarbon	>35	<10	High % of C2–C7	>30%PV	Sandstone or carbonate	Thin unless dipping	N.C.	>2,000 (LPG) to >5,000 (H.P. Gas)	N.C.
Nitrogen & flue gas	>24, >35 for N2	<10	High % of C1–C7	>30% PV	Sandstone or carbonate	Thin unless dipping	N.C.	>4,500	N.C.
Carbon dioxide	>26	<15	High % of C2–C12	>30% PV	Sandstone or carbonate	Thin unless dipping	N.C.	>2,000	N.C.
Chemical Flooding									
Surfactant/polymer	>25	<30	Light intermediates desired	>30% PV	Sandstone preferred	>10	>20	<8,000	<175
Polymer	>25	<150	N.C.	>10% PV Mobile oil	Sandstone preferred; carbonate possible	N.C.	>10 (normally)	<9,000	<200
Alkaline	13–35	<200	Some organic acids	Above waterflood residual	Sandstone preferred	N.C.	>20	<9,000	<200
Thermal									
Combustion	<40 (10–25 normally)	<1,000	Some asphaltic components	>40–50% PV	Sand or sand stone with high porosity	>10	>100*	>500	>150 preferred
Steamflooding	<25	>20	N.C.	>40–50% PV	Sand or sand stone with high porosity	>20	>200**	300–5,000	N.C.

From Reference 12.
N.C. = Not Critical.
*Transmissibility >20 md ft/cp
**Transmissibility >100 md ft/cp

simulations to predict recovery, and the performance of the project. One of the first steps in deciding to consider EOR is, of course, to select reservoirs with sufficient recoverable oil and areal extent to make the venture profitable.

With any of the processes, the nature of the reservoir will play a dominant role in the success or failure of the process. Many of the failures with EOR have resulted because of unknown or unexpected reservoir problems. Thus, a thorough geological study is usually warranted.

The technique of using cursory screening guides is convenient for gaining a quick overview of all possible methods before selecting the best one for an economic analysis. Common sense and caution must be exercised since the technical guides are based on laboratory data and results of enhanced recovery field trials, and are not rigid guides for applying certain processes to specific reservoirs. Additionally, the technical merits of recent field projects are clouded by various incentive programs that make it difficult to discern true technical applications. Some projects may have been technical misapplications or failures, but economic successes. Certainly, there have been enough technical successes, but economic failures.

Nevertheless, some EOR processes can be rejected quickly because of unfavorable reservoir or oil properties, so the use of preferred criteria can be helpful in selecting methods that may be commercially attractive. If the criteria are too restrictive, some feasible method may be rejected from consideration. Therefore, the guidelines that are adopted should be sufficiently broad to encompass essentially all of the potential methods for a candidate reservoir.

For convenience, brief descriptions of the eight most common enhanced recovery methods are provided in the following sections. These descriptions list the salient features of each method along with the important screening guides. A few general comments are offered here on the relative importance of some individual screening guides to the overall success of the various methods. In addition, we will make some observations on the method itself and its relationship to other enhanced recovery choices that may be available.

5.5.1 Hydrocarbon Miscible Flooding [12]

Description Hydrocarbon miscible flooding consists of injecting light hydrocarbons through the reservoir to form a miscible flood. Three different methods are used. One method uses about 5% PV slug of liquidified petroleum gas (LPG) such as propane, followed by natural gas or gas and water. A second method, called enriched (condensing) gas drive, consists of injecting a 10%–20% PV slug of natural gas that is enriched with ethane through hexane (C_2 to C_6), followed by lean gas (dry, mostly methane) and possibly water. The enriching components are transferred from the gas to the oil. The third method, called high pressure (vaporizing) gas drive, consists of injecting lean gas at high pressure to vaporize $C_2 - C_6$ components from the crude oil being displaced.

Mechanisms   Hydrocarbon miscible flooding recovers crude oil by:
- Generating miscibility (in the condensing and vaporizing gas drive)
- Increasing the oil volume (swelling)
- Decreasing the viscosity of the oil

Technical Screening Guides
Crude oil

Gravity	> 35° API
Viscosity	< 10 cp
Composition	High percentage of light hydrocarbons (C_2-C_7)

Reservoir

Oil saturation	> 30% PV
Type of formation	Sandstone or carbonate with a minimum of fractures and high permeability streaks
Net thickness	Relatively thin unless formation is steeply dipping
Average permeability	Not critical if uniform
Depth	> 2,000 ft (LPG) to > 5000 ft (high pressure gas)
Temperature	Not critical

Limitations
- The minimum depth is set by the pressure needed to maintain the generated miscibility. The required pressure ranges from about 1,200 psi for the LPG process to 3,000–5,000 psi for the high pressure gas drive, depending on the oil.
- A steeply dipping formation is very desirable to permit some gravity stabilization of the displacement which normally has an unfavorable mobility ratio.

Problems
- Viscous fingering results in poor vertical and horizontal sweep efficiency.
- Large quantities of expensive products are required.
- Solvent may be trapped and not recovered.

5.5.2 Nitrogen and Flue Gas Flooding [12]

Description   Nitrogen and flue gas flooding are oil recovery methods which use these inexpensive nonhydrocarbon gases to displace oil in systems which may be either miscible or immiscible depending on the pressure and oil composition. Because of their low cost, large volumes of these gases may be injected. Nitrogen or flue gas are also considered for use as chase gases in hydrocarbon-miscible and CO_2 floods.

Mechanisms   Nitrogen and flue gas flooding recover oil by:
- Vaporizing the lighter components of the crude oil and generating miscibility if the pressure is high enough.
- Providing a gas drive where a significant portion of the reservoir volume is filled with low-cost gases.

Technical Screening Guides
Crude oil

Gravity	> 24° API(> 35° for nitrogen)
Viscosity	< 10 cp
Composition	High percentage of light hydrocarbons (C_1–C_7)

Reservoir

Oil saturation	> 30% PV
Type of formation	Sandstone or carbonate with few fractures and high permeability streaks
Net thickness	Relatively thin unless formation is dipping
Average permeability	Not critical
Depth	>4.500 ft
Temperature	Not critical

Limitations
- Developed miscibility can only be achieved with light oils and at high pressures; therefore, deep reservoirs are needed.
- A steeply dipping formation is desired to permit gravity stabilization of the displacement which has a very unfavorable mobility ratio.

Problems
- Viscous fingering results in poor vertical and horizontal sweep efficiency.
- Corrosion can cause problems in the flue gas method.
- The nonhydrocarbon gases must be separated from the saleable produced gas.

5.5.3   Carbon Dioxide Flooding [12]

Description   Carbon dioxide flooding is carried out by injecting large quantities of CO_2 (15% or more of the hydrocarbon PV) into the reservoir. Although CO_2 is not truly miscible with the crude oil, the CO_2 extracts the light-to-intermediate components from the oil, and, if the pressure is high enough, develops miscibility to displace the crude oil from the reservoir.

Mechanisms CO_2 recovers crude oil by:
- Generation of miscibility
- Swelling the crude oil
- Lowering the viscosity of the oil
- Lowering the interfacial tension between the oil and the CO_2-oil phase in the near-miscible regions.

Technical Screening Guides
Crude oil

Gravity	$> 26°API$ (preferably $> 30°$)
Viscosity	$< 15\,cp$ (preferably $< 10\,cp$)
Composition	High percentage of intermediate hydrocarbons $(C_5–C_{20})$, especially $C_5–C_{12}$

Reservoir

Oil saturation	$> 30\%$ PV
Type of formation	Sandstone or carbonate with a minimum of fractures and high permeability streaks
Net thickness	Relatively thin unless formation is steeply dipping.
Average permeability	Not critical if sufficient injection rates can be maintained.
Depth	Deep enough to allow high enough pressure ($>$ about 2,000 ft), pressure required for optimum production (sometime called minimum miscibility pressure) ranges from about 1,200 psi for a high gravity ($> 30°$ API) crude at low temperatures to over 4,500 psi for heavy crudes at higher temperatures.
Temperature	Not critical but pressure required increases with temperature.

Limitations
- Very low viscosity of CO_2 results in poor mobility control.
- Availability of CO_2.

Problems
- Early breakthrough of CO_2 causes several problems: corrosion in the producing wells; the necessity of separating CO_2 from saleable hydrocarbons; repressuring of CO_2 for recycling; and a high requirement of CO_2 per incremental barrel produced.

5.5.4 Surfactant/Polymer Flooding

Description Surfactant/polymer flooding, also called micellar/polymer or microemulsion flooding, consists of injecting a slug that contains water, surfactant, electrolyte (salt), usually a cosolvent (alcohol), and possibly a hydrocarbon (oil). The size of the slug is often 5%–15% PV for a high surfactant concentration system and 15%–50% PV for low concentrations. The surfactant slug is followed by polymer-thickened water. Concentrations of the polymer often range from 500–2,000 mg/L; the volume of polymer solution injected may be 50% PV, more or less, depending on the process design.

Mechanisms Surfactant/polymer flooding recovers crude oil by:
- Lowering the interfacial tension between oil and water
- Solubilization of oil
- Emulsification of oil and water
- Mobility enhancement

Technical Screening Guides
Crude oil

Gravity	> 25° API
Viscosity	< 30 cp
Composition	Light intermediates are desirable

Reservoir

Oil saturation	> 30% PV
Type of formation	Sandstone preferred
Net thickness	> 10 ft
Average permeability	> 20 md
Depth	< about 8,000 ft (see temperature)
Temperature	< 175°F

Limitations
- An areal sweep of more than 50% on waterflood is desired.
- Relatively homogeneous formation is preferred.
- High amounts of anhydrite, gypsum, or clays are undesirable.
- Available systems provide optimum behavior over a very narrow set of conditions.
- With commercially available surfactants, formation water chlorides should be < 20,000 ppm and divalent ion (Ca^{++} and Mg^{++}) < 500 ppm.

Problems
- Complex and expensive system.
- Possibility of chromatographic separation of chemicals.
- High adsorption of surfactant.

- Interactions between surfactant and polymer.
- Degradation of chemicals at high temperature.

5.5.5 Polymer Flooding [12]

Description   The objective of polymer flooding is to provide better displacement and volumetric sweep efficiencies during a waterflood. Polymer augmented waterflooding consists of adding water soluble polymers to the water before it is injected into the reservoir. Low concentrations (often 250–2,000 mg/L) of certain synthetic or biopolymers are used; properly sized treatments may require 15%–25% reservoir PV.

Mechanisms   Polymers improve recovery by:
- Increasing the viscosity of water
- Decreasing the mobility of water
- Contacting a large volume of the reservoir

Technical Screening Guides
Crude oil

Gravity	> 25° API
Viscosity	< 150 cp (preferably < 100)
Composition	Not critical

Reservoir

Oil saturation	> 10% PV mobile oil
Type of formation	Sandstones preferred but can be used in carbonates
Net thickness	Not critical
Average permeability	> 10 md (as low as 3 md in some cases)
Depth	< about 9,000 ft (see temperature)
Temperature	< 200°F to minimize degradation

Limitations
- If oil viscosities are high, a higher polymer concentration is needed to achieve the desired mobility control.
- Results are normally better if the polymer flood is started before the water-oil ratio becomes excessively high.
- Clays increase polymer adsorption.
- Some heterogeneities are acceptable, but for conventional polymer flooding, reservoirs with extensive fractures should be avoided. If fractures are present, the crosslinked or gelled polymer techniques may be applicable.

Problems
- Lower injectivity than with water can adversely affect oil production rate in the early stages of the polymer flood.

- Acrylamide-type polymers lose viscosity due to shear degradation or increases in salinity and divalent ions.
- Xanthan gum polymers cost more, are subject to microbial degradation, and have a greater potential for wellbore plugging.

5.5.6 Alkaline Flooding [12]

Description Alkaline or caustic flooding involves the injection of chemicals such as sodium hydroxide, sodium silicate or sodium carbonate. These chemicals react with organic petroleum acids in certain crudes to create surfactants in situ. They also react with reservoir rocks to change wettability. The concentration of the alkaline agent is normally 0.2 to 5%; slug size is often 10% to 50% PV, although one successful flood only used 2% PV, (but this project also included polymers for mobility control). Polymers may be added to the alkaline mixture, and polymer-thickened water can be used following the caustic slug.

Mechanisms Alkaline flooding recovers crude oil by:
- A reduction of interfacial tension resulting from the produced surfactants
- Changing wettability from oil-wet to water-wet
- Changing wettability from water-wet to oil-wet
- Emulsification and entrainment of oil
- Emulsification and entrapment of oil to aid in mobility control
- Solubilization of rigid oil films at oil-water interfaces (Not all mechanisms are operative in each reservoir.)

Technical Screening Guides
Crude oil

Gravity	> 13° to 35° API
Viscosity	< 200 cp
Composition	Some organic acids required

Reservoir

Oil saturation	Above waterflood residual
Type of formation	Sandstones preferred
Net thickness	Not critical
Average permeability	> 20 md
Depth	< about 9,000 ft (see temperature)
Temperature	< 200°F preferred

Limitations
- Best results are obtained if the alkaline material reacts with the crude oil: the oil should have an acid number of more than 0.2 mg KOH/g of oil.
- The interfacial tension between the alkaline solution and the crude oil should be less than 0.01 dyne/cm.

- At high temperatures and in some chemical environments, excessive amounts of alkaline chemicals may be consumed by reaction with clays, minerals, or silica in the sandstone reservoir.
- Carbonates are usually avoided because they often contain anhydrite or gypsum, which interact adversely with the caustic chemical.

Problems
- Scaling and plugging in the producing wells.
- High caustic consumption

5.5.7 In-Situ Combustion [12]

Description   In-situ combustion or fireflooding involves starting a fire in the reservoir and injecting air to sustain the burning of some of the crude oil. The most common technique is forward combustion in which the reservoir is ignited in an injection well, and air is injected to propagate the combustion front away from the well. One of the variations of this technique is a combination of forward combustion and waterflooding (COFCAW). A second technique is reverse combustion in which a fire is started in a well that will eventually become a producing well, and air injection is then switched to adjacent wells; however, no successful field trials have been completed for reverse combustion.

Mechanisms   In-situ combustion recovers crude oil by:
- The application of heat which is transferred downstream by conduction and convection, thus lowering the viscosity of the crude.
- The products of steam distillation and thermal cracking which are carried forward to mix with and upgrade the crude.
- Burning coke that is produced from the heavy ends of the crude oil.
- The pressure supplied to the reservoir by the injected air.

Technical Screening Guides
Crude oil

Gravity	$< 40°$ API (normally 10–$25°$)
Viscosity	$< 1,000$ cp
Composition	Some asphaltic components to aid coke deposition

Reservoir

Oil saturation	> 500 bbl/acre-ft (or > 40–50% PV)
Type of formation	Sand or sandstone with high porosity
Net thickness	> 10 ft
Average permeability	> 100 md
Transmissibility	> 20 mud ft/cp
Depth	> 500 ft
Temperature	$> 150°$F preferred

Limitations

- If sufficient coke is not deposited from the oil being burned, the combustion process will not be sustained.
- If excessive coke is deposited, the rate of advance of the combustion zone will be slow, and the quantity of air required to sustain combustion will be high.
- Oil saturation and porosity must be high to minimize heat loss to rock.
- Process tends to sweep through upper part of reservoir so that sweep efficiency is poor in thick formations.

Problems

- Adverse mobility ratio.
- Complex process, requiring large capital investment, is difficult to control.
- Produced flue gases can present environmental problems.
- Operational problems such as severe corrosion caused by low pH hot water, serious oil-water emulsions, increased sand production, deposition of carbon or wax, and pipe failures in the producing wells as a result of the very high temperatures.

5.5.8 Steamflooding [12]

Description The steam drive process or steamflooding involves the continuous injection of about 80% quality steam to displace crude oil towards producing wells. Normal practice is to precede and accompany the steam drive by a cyclic steam stimulation of the producing wells (called huff and puff).

Mechanisms Steam recovers crude oil by:
- Heating the crude oil and reducing its viscosity
- Supplying pressure to drive oil to the producing well

Technical Screening Guides
Crude oil

Gravity	< 25° API (normal range is 10°–25° API)
Viscosity	> 20 cp (normal range is 100–5,000 cp)
Composition	Not critical but some light ends for stream distillation will help

Reservoir

Oil saturation	> 500 bbl/acre-ft (or > 40%–50% PV)
Type of formation	Sand or sandstone with high porosity and permeability preferred
Net thickness	> 20 feet

Average permeability	> 200 md (see transmissibility)
Transmissibility	> 100 md ft/cp
Depth	300–5,000 ft
Temperature	Not critical

Limitations

- Oil saturations must be quite high and the pay zone should be more than 20 feet thick to minimize heat losses to adjacent formations.
- Lighter, less viscous crude oils can be steamflooded but normally will not be if the reservoir will respond to an ordinary waterflood.
- Steamflooding is primarily applicable to viscous oils in massive, high permeability sandstones or unconsolidated sands.
- Because of excessive heat losses in the wellbore, steamflooded reservoirs should be as shallow as possible as long as pressure for sufficient injection rates can be maintained.
- Steamflooding is not normally used in carbonate reservoirs.
- Since about one-third of the additional oil recovered is consumed to generate the required steam, the cost per incremental barrel of oil is high.
- A low percentage of water-sensitive clays is desired for good injectivity.

Problems

- Adverse mobility ratio and channeling of steam.

5.5.9 Criteria for Gas Injection

For LPG slug or solvent flooding, enriched (condensing) gas drive, and high pressure (vaporizing) gas drive, a range of pressures (and therefore, depths) are needed to achieve miscibility in the systems. Thus, there is a minimum depth requirement for each of the processes as shown earlier (see section on "Hydrocarbon Miscible Flooding"). The permeability is not critical if the structure is relatively uniform; permeabilities of the reservoirs for the current field projects range from less than 1 md to several darcies [29]. On the other hand, the crude oil characteristics are very important. A high-gravity, low-viscosity oil with a high percentage of the C_2–C_7 intermediates is essential if miscibility is to be achieved in the vaporizing gas drives.

As shown earlier under "Nitrogen and Flue Gas Flooding," the screening criteria for flooding with nitrogen or flue gas are similar to those for the high pressure gas drive. Pressure and depth requirements, as well as the need for a very light oil, are even greater if full miscibility is to be realized in the reservoir. The nitrogen and flue gas method is placed between hydrocarbon miscible and CO_2 flooding because the process can also recover oil in the immiscible mode. It can be economic because much of the reservoir space is filled with low cost gas.

Because of the minimum pressure requirement, depth is an important screening criteria, and CO_2 floods are normally carried out in reservoirs that are more than 2,000 ft deep. The oil composition is also important (see section on "Carbon Dioxide Flooding"), and the API gravity exceeds 30° for most of the active CO_2 floods [29]. A notable exception is the Lick Creek, Arkansas, CO_2/waterfloood project which was conducted successfully, not as a miscible project, but as an immiscible displacement [30].

5.5.10 Criteria for Chemical Methods

For surfactant/polymer methods, oil viscosities of less than 30 cp are desired so that adequate mobility control can be achieved. Good mobility control is essential for this method to make maximum utilization of the expensive chemicals. Oil saturations remaining after a waterflood should be more than 30% PV to ensure that sufficient oil is available for recovery. Sandstones are preferred because carbonate reservoirs are heterogeneous, contain brines with high divalent ion contents, and cause high adsorption of commonly used surfactants. To ensure adequate injectivity, permeability should be greater than 20 md. Reservoir temperature should be less than 175°F to minimize degradation of the presently available surfactants. A number of other limitations and problems were mentioned earlier, including the general requirement for low salinity and hardness for most of the commercially available systems. Obviously, this method is very complex, expensive, and subject to a wide range of problems. Most importantly, the available systems provide optimum reduction in interfacial tension over a very narrow salinity range. Preflushes have been used to attempt to provide optimum conditions, but they have often been ineffective.

The screening guidelines and a description of polymer flooding are contained earlier in Section "Polymer Flooding." Since the objective of polymer flooding is to improve the mobility ratio without necessarily making the ratio favorable, the maximum oil viscosity for this method is 100 or possibly 150 cp. If oil viscosities are very high, higher polymer concentrations are needed to achieve the desired mobility control, and thermal methods may be more attractive. As discussed earlier, polymer flooding will not ordinarily mobilize oil that has been completely trapped by water; therefore, a mobile oil saturation of more than 10% is desired. In fact, a polymer flood is normally more effective when started at low producing water-oil ratios [31]. Although sandstone reservoirs are usually preferred, several large polymer floods have been conducted in carbonate reservoirs. Lower-molecular-weight polymers can be used in reservoirs with permeabilities as low as 10 md (and, in some carbonates, as low as 3 md). While it is possible to manufacture even lower-molecular-weight polymers to inject into lower permeability formations, the amount of viscosity generated per pound of polymer would not be enough to make such products of interest.

With current polymers, reservoir temperature should be less than 200°F to minimize degradation; this requirement limits depths to about 9,000 ft. A potentially serious problem with polymer flooding is the decrease the injectivity which must accompany any increase in injection fluid viscosity. If the decreased injectivity is prolonged, oil production rates and project costs can be adversely affected. Injection rates for polymer solutions may be only 40%–60% of those for water alone, and the reduced injectivity may add several million dollars to the total project costs. Other problems common to the commercial polymers are cited earlier.

Moderately low gravity oils (13°–35° API) are normally the target for alkaline flooding (see section on "Alkaline Flooding"). These oils are heavy enough to contain the organic acids, but light enough to permit some degree of mobility control. The upper viscosity limit (< 200 cp) is slightly higher than for polymer flooding. Some mobile oil saturation is desired, the higher the better. The minimum average permeability is about the same as for surfactant/polymer (> 20 md). Sandstone reservoirs are preferred since carbonate formations often contain anhydrite or gypsum which react and consume the alkaline chemicals. The alkaline materials also are consumed by clays, minerals, or silica; this consumption is high at elevated temperatures so the maximum desired temperature is 200°F. Caustic consumption in field projects has been higher than indicated by laboratory tests. Another potential problem in field applications is scale formation which can result in plugging in the producing wells.

5.5.11 Criteria for Thermal Methods

For screening purposes, steamflooding and fireflooding are often considered together. In general, combustion should be the choice when heat losses from steamflooding would be too great. In other words, combustion can be carried out in deeper reservoirs and thinner, tighter sand sections where heat losses for steamflooding are excessive. Screening guides for in-situ combustion are given earlier in Section "In-Situ Combustion." The ability to inject at high pressures is usually important so 500 ft has been retained as the minimum depth, but a few projects have been done at depths of less than 500 ft. Since the fuel and air consumption decrease with higher gravity oils, there is a tendency to try combustion in lighter oils if the fire can be maintained, but no projects have been done in reservoirs with oil gravities greater than 32° API [29].

In summary, if all screening criteria are favorable, fireflooding appears to be an attractive method for reservoirs that cannot be produced by methods used for the lighter oils. However, the process is very complicated and beset with many practical problems such as corrosion, erosion and poorer mobility ratios than steamflooding. Therefore, when the economics are comparable, steam injection is preferred to a combustion drive [4].

Screening criteria for steamflooding are listed earlier in section "Steam-flooding". Although steamflooding is commonly used with oils ranging in gravity from 10°–25° API, some gravities have been lower, and there is recent interest in steamflooding light oil reservoirs. Oils with viscosities of less than 20 cp are usually not candidates for steamflooding because waterflooding is less expensive; the normal range is 100–5,000 cp. A high saturation of oil-in-place is required because of the intensive use of energy in the generation of steam. In order to minimize the amount of rock heated and maximize the amount of oil heated, formations with high porosity are desired; this means that sandstones or unconsolidated sands are the primary target, although a steam drive pilot has been conducted in a highly fractured carbonate reservoir in France. The product of oil saturation times porosity should be greater than about 0.08 [26]. The fraction of heat lost to the cap and base rocks varies inversely with reservoir thickness. Therefore, the greater the thickness of the reservoir, the greater the thermal efficiency. Steamflooding is possible in thin formations if the permeability is high. High permeabilities (> 200 md or preferably > 500 md) are needed to permit adequate steam injectivity; transmissibility should be greater than 100 md ft/cp at reservoir conditions. Depths shallower than about 300 ft may not permit good injectivity because the pressures required may exceed fracture gradients. Heat losses become important at depths greater than about 2,500 ft. and steamflooding is not often considered at depths greater than 5,000 ft. Downhole steam generators may have potential in deeper formations if operational problems can be overcome.

5.5.12 Graphical Representation of Screening Guides

From the summary of screening guides in Table 5.1, the viscosity, depth and permeability criteria are presented graphically in Figures 5.3 to 5.5. The figures have some features which permit the quick application of screening criteria but they cannot replace the table for detailed evaluations. In a sense, the figures present a truer picture than the table because there are few absolutes among the numbers presented as screening guides in the tables. Different authors and organizations may use different parameters for the same process, and most of the guidelines are subject to change as new laboratory and field information evolves. In field applications, there are exceptions to some of the accepted criteria, and the graphs accommodate these nicely. The "greater than" and "less than" designations of the tables can also be displayed better graphically.

The range of values are indicated on the graphs by the open areas, and by cross-hatching along with general words such as "more difficult," "not feasible," etc. The "good" or "fair" ranges are those usually encompassed by the screening parameters in the table. However, the notation of "good" or "very good" does not mean that the indicated process is sure to

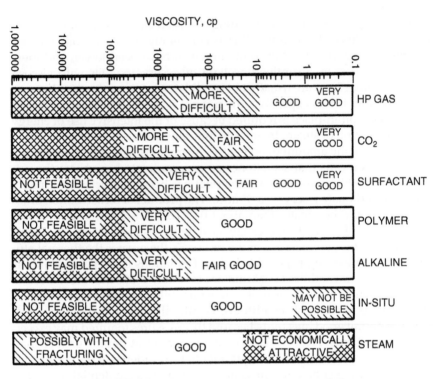

FIGURE 5.3   Viscosity ranges for EOR processes [12].

work; it means simply that it is in the preferred range for that oil or reservoir characteristic.

The influence of viscosity on the technical feasibility of different enhanced recovery methods is illustrated in Figure 5.3. Note the steady progression, with increasing viscosity, from those processes that work well with very light oils (hydrocarbon miscible or nitrogen) to oils that are so viscous that no recovery is possible unless mining and extraction are employed.

For completeness, we have included the two "last resort" methods (special steamflooding techniques with shafts, fractures, drainholes, etc., and mining plus extraction) are listed in Figure 5.5. These methods are not included in Figures 5.4 and 5.5 because these unconventional techniques are not considered in most reservoir studies.

Figure 5.4 shows that those enhanced recovery processes that work well with light oils have rather specific depth requirements. As discussed, each gas injection method has a minimum miscibility pressure for any given oil, and the reservoir must be deep enough to accommodate the required pressure.

Figure 5.5 shows that the three methods that rely on gas injection are the only ones that are even technically feasible at extremely low permeabilities.

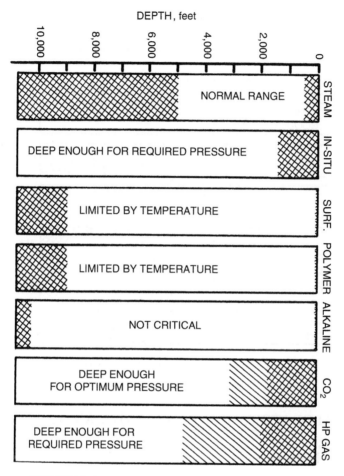

FIGURE 5.4   Depth requirements for EOR processes [12].

The three methods that use backup waterflooding need a permeability of greater than 10 md in order to inject the chemicals or emulsions and to produce the released oil from the rock. Although most authors show a minimum permeability requirement of 20 md for polymers, we indicate a possible range down as low as 3 md for low molecular weight polymers, especially in some carbonate reservoirs.

The screening guides in the figures can perhaps be summarized by stating a fact well-known to petroleum engineers: oil recovery is easiest with light oil in very permeable reservoirs and at shallow or intermediate depths. Unfortunately, nature has not been kind in the distribution of hydrocarbons, and it is necessary to select the recovery method that best matches the oil and reservoir characteristics.

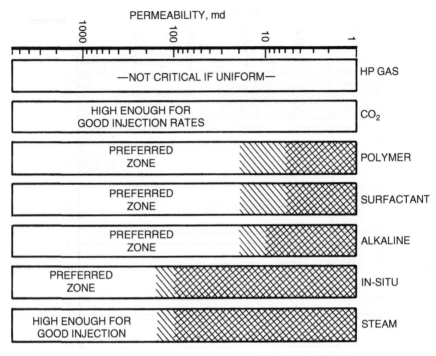

FIGURE 5.5   Permeability ranges for EOR methods [12].

5.6 LABORATORY DESIGN FOR ENHANCED RECOVERY

5.6.1 Preliminary Tests

Water Analysis   A complete water analysis is important to determine the effects of dissolved ions on the EOR processes (especially the chemical methods) or to ascertain any potential water problems such as scale or corrosion that may result when EOR processes are implemented. Water viscosity and density are also measured.

Oil Analysis   Oil viscosity and density are measured as well. A carbon number distribution of the crude may be obtained, especially if CO_2 flooding is being considered.

Core Testing   Routine core analyses, such as porosity, permeability, relative permeabilities, capillary pressure, and waterflood susceptibility tests are normally done by service companies that specialize in these types of tests. Specialized core tests, such as thin sections or scanning electron microscopy, are available to evaluate the relationship between pore structure and the process being considered. If required, stimulation or injectivity improvement measures can be recommended.

5.6.2 Polymer Testing

The desirability of adding polymers is determined by evaluating all available data to assess the performance of normal waterflooding. Any problems such as adverse viscosity ratios or large permeability variations should be identified. If the results of this study indicate that mobility control of the waterflood is warranted, the following laboratory tests are undertaken.

Viscosity Testing   Based on the permeability of the reservoir, relative permeability data, and the desired level of mobility control, polymers of certain molecular weights are selected for testing. Various concentrations of the polymers are dissolved in both the available injection water and in blends of the injection and formation waters. Polymer solutions may be non-Newtonian at certain shear rates, that is, the viscosity decreases at high shear rates (shear-thinning or pseudoplastic). This shear-thinning behavior is reversible and, if observed in the reservoir, is beneficial in that good injectivity can result from the lower viscosity observed at high shear rates near the injection well. At the lower shear rates encountered some distance from the injector, the polymer solution develops a higher viscosity. In this testing, it is important to consider not only the viscosity of the injected solution, but, more importantly, the in-situ viscosity that is achieved in the reservoir. Several things can happen that will reduce viscosity when the polymer solution is injected into the reservoir. Reduction in viscosity as a result of irreversible shear degradation is possible at the injection wellbore if the shear rates or shear stresses are large. Once in the reservoir, dilution with formation water or ion exchange with reservoir minerals can cause a reduction in viscosity, and the injected polymer concentration will need to be sufficiently high to compensate for all viscosity-reducing effects.

Polymer Retention   Retention of polymer in a reservoir can result from adsorption, entrapment, or, with improper application, physical plugging. Polymer retention tests are usually performed after a standard waterflood (at residual oil saturation) or during a polymer flood oil recovery test. If polymer retention tests are conducted with only water initially present in the core, a higher level of retention will result from the increased surface area available to the polymer solution in the absence of oil. Effluent samples from the core are collected during both the polymer injection and a subsequent water flush. These samples are analyzed for polymer content. From a material balance, the amount of polymer retained in the core is calculated. Results are usually expressed in lbs per acre-ft. Excessive retention will increase the amount of polymer that must be added to achieve the desired mobility control. The level of polymer retained in a reservoir depends on a number of variables: permeability of the rock, surface area, nature of the reservoir rock (sandstone, carbonate, minerals, or clays), nature of the solvent for the polymer (salinity and hardness), molecular weight of the polymer, ionic charge on the polymer, and the volume of porosity that is not accessible to the flow of polymer solution. Polymer retention levels often range from less than 100 lb/acre-ft to several hundred lb/acre-ft.

5.6.3 Surfactant and Alkali Testing

Laboratory tests consist of measuring the interfacial tension (IFT) between the crude oil and the injected solution (alkaline or surfactant additive). This is usually done with a spinning drop interfacial tensiometer. With surfactants, the requirement for measuring tensions can be minimized by performing vial tests to determine solubilization parameters that can be correlated with IFT. Other tests include determining relative permeabilities, wettability, and total fluid mobilites. Once the optimum conditions are found, results of oil recovery tests with the chemical flood additives are conducted, usually at waterflood residual oil saturation.

5.6.4 CO_2 Flooding

For the gas injection projects, the trend in this country is toward the use of carbon dioxide although the full impact of CO_2 flooding will be felt in several years since construction of CO_2 pipelines into the west Texas area was completed in the 1980s. Carbon dioxide flooding is not a truly miscible process; that is, it does not dissolve in all proportions with crude oil. However, CO_2 can extract light to intermediate components out of the crude oil. This CO_2-rich mixture can develop miscibility and effectively displace additional crude oil. The main limitation involved is the very low viscosity of CO_2 that results in fingering of CO_2 through the more viscous crude oil. This causes premature breakthrough of the CO_2 and reduces the amount of oil recovered per unit volume of CO_2.

A prediction of the minimum pressure required to achieve miscibility can be made if the reservoir depth and basic properties of the crude oil are known. Laboratory tests often consist of some means of determining the minimum miscibility pressure, often by observing the oil displacement efficiency by CO_2 in a small-diameter tube (slim tube) packed with sand or glass beads. Carbon number distribution of the crude will be of value in determining if sufficient amounts of the C_5 to C_{12} components are present.

5.6.5 Thermal Recovery

Viscosities of very viscous crude oils can be reduced by the use of thermal recovery methods. Fireflooding or in-situ combustion involves starting a fire in the reservoir and injecting air to sustain the burning of some of the crude oil. Heat that is generated lowers the viscosity of the crude oil and results in improved recovery. With the steam drive or steamflooding process, steam is generated on the surface and injected into the injection wells. Some companies are now exploring the use of downhole steam generators in deeper wells where heat loss can be a serious problem. A primary problem with steam flooding is the channeling of steam through thin sections of the reservoir. To combat this problem, several organizations are studying the use of surfactants to create a foam in situ for improving sweep efficiency.

For steamflooding, the most important laboratory tests are, of course, viscosity of the crude oil and permeability of the reservoir core material. To be economically viable, steamfloods must be conducted in thick, very permeable, shallow reservoirs that contain very viscous crude.

References

[1] van Poollen, H. K., and Associates, Inc., *Fundamentals of Enhanced Oil Recovery*, PenWell Publishing Co., Tulsa (1980).

[2] Stalkup, F. I., Jr., *Miscible Displacement*, SPE Monograph Series, SPE, Dallas, Vol. 8 (1983).

[3] Klins, M. A., *Carbon Dioxide Flooding*. International Human Resources Development Corp., Boston (1984).

[4] Prats, M., *Thermal Recovery*, Monograph Series, SPE, Dallas, Vol. 7(1982).

[5] Latil, M., et al., *Enhanced Oil Recovery*, Gulf Publishing Co., Houston (1980).

[6] Haynes, H. J., et al., *Enhanced Oil Recovery*, National Petroleum Council; Industry Advisory Council to the U.S. Department of the Interior (1976).

[7] Bailey, R. E., et al., *Enhanced Oil Recovery*, Natl. Petroleum Council, Washington (June 1984).

[8] Broome, J. H., Bohannon, J. M., and Stewart, W. C., "The 1984 Natl. Petroleum Council Study on EOR: An Overview," *J. Pet. Tech.* (Aug. 1986), pp. 869–874.

[9] Doe, P. H., Carey, B. S., and Helmuth, E. S., "The Natl. Petroleum Council EOR Study: Chemical Processes," paper SPE 13240 presented at the 1984 Annual Tech. Conf. & Exhib., Sept. 16–19.

[10] King, J. E., Blevins, T. R., and Britton, M. W., "The National Petroleum Council EOR Study: Thermal Processes," paper SPE 13242 presented at the 1984 SPE Annual Technical Conf. and Exhib., Houston, Sept. 16–19.

[11] Robl, F. W., Emanuel, A. S., and Van Meter, O. E., Jr., "The 1984 Natl. Petroleum Council Estimate of Potential of EOR for Miscible Processes," *J. Pet. Tech.* (Aug. 1986), pp. 875–882.

[12] Taber, J. J., and Martin, F. D., "Technical Screening Guides for the Enhanced Recovery of Oil," paper SPE 12069 presented at the SPE 1983 Annual Technical Conf. & Exhib., San Francisco, October 5–8.

[13] Chang, H. L., "Polymer Flooding Technology—Yesterday, Today and Tomorrow," *J. Pet. Tech.* (Aug. 1978), pp. 1113–1128.

[14] Martin, F. D., "Design and Implementation of a Polymer Flood," Southwestern Petroleum Short Course, *Proc.*, 33rd Annual Southwestern Petroleum Short Course, Lubbock, April 23–24, 1986.

[15] Taber, J. J., "The Injection of Detergent Slugs in Water Floods," *Trans.* AIME, Vol. 213 (1958), pp. 186–192.

[16] Gogarty, W. B., "Status of Surfactant or Micellar Methods," *J. Pet. Tech.* (Jan. 1976), pp. 93–102.

[17] Gogarty, W. B., "Micellar/Polymer Flooding—An Overview," *J. Pet. Tech.* (Aug. 1978), pp. 1089–1101.

[18] Gogarty, W. B., "Enhanced Oil Recovery Through the Use of Chemicals—Part 1," *J. Pet. Tech* (Sept. 1983), pp. 1581–1590.

[19] Gogarty, W. B., "Enhanced Oil Recovery Through the Use of Chemicals—Part 2," *J. Pet. Tech.* (Oct. 1983), pp. 1767–1775.

[20] Taber, J. J., "Enhanced Recovery Methods for Heavy and Light Oils," *Proc.*, International Conference on Heavy Versus Light Oils: Technical Issues and Economic Considerations, Colorado Springs, March 24–26, 1982.

[21] Holm, L. W., and Josendal, V. A., "Effect of Oil Composition on Miscible Type Displacement by Carbon Dioxide," *Soc. Pet. Eng. J.* (Feb. 1982), pp. 87–98.

[22] Orr, F. M., Jr., and Jensen, C. M., "Interpretation of Pressure-Composition Phase Diagrams for CO_2-Crude Oil Systems," paper SPE 11125 presented at SPE 57th Annual Fall Technical Conf. and Exhibition, New Orleans, Sept. 26–29, 1982.

[23] Helier, J. P., and Taber, J. J., "Influence of Reservoir Depth on Enhanced Oil Recovery by CO_2 Flooding," paper SPE 15001 presented at the 1986 SPE Permian Basin Oil & Gas Recovery Conference, Midland, March 13–14.

[24] Chu, C., "Current In-Situ Combustion Technology," *J. Pet. Tech.* (Aug. 1983), pp. 1412–1418.

[25] Chu, C., "State-of-the-Art Review of Fireflood Field Projects," *J. Pet. Tech.* (Jan. 1982), pp. 19–36.

[26] Chu, C., "State-of-the-Art Review of Steamflood Field Projects," *J. Pet. Tech.* (Oct. 1985), pp. 1887–1902.

[27] White, P. D., "In-Situ Combustion Appraisal and Status," *J. Pet. Tech.* (Nov. 1985), pp. 1943–1949.

[28] Garon, A. M., Kumar, M., and Cala, G. C., "The State of the Art Oxygen Fireflooding," *In Situ*, Vol. 10, No. 1 (1986), pp. 1–26.

[29] Leonard, J., "Increased Rate of EOR Brightens Outlook," *Oil & Gas J.* (April 14, 1986), pp. 71–101.

[30] Reid, R. B., and Robinson, J. J., "Lick Creek Meakin Sand Unit Immiscible CO_2/Waterflood Project," *J. Pet. Tech.* (Sept. 1981), pp. 1723–1729.

[31] Agnew, H. J., "Here's How 56 Polymer Oil-Recovery Projects Shape Up," *Oil & Gas J.* (May 1972), pp. 109–112.

[32] *Improved Oil Recovery*, Interstate Oil Compact Commission, Oklahoma City (March 1983).

Index

Notes: Page numbers followed by "f" refer to figures; page numbers followed by "t" refer to tables.

Printed and bound by CPI Group (UK) Ltd, Croydon, CR0 4YY

03/10/2024

01040435-0006